The presentation of Lange's first watch collection of the modern era on October 24, 1994 will always remain for me—and certainly not only me—unforgettable. Round about 150 years after the remarkable initiative of watchmaker Ferdinand Adolph Lange, a new generation of entrepreneurs had stepped up to the plate to fill an almost forgotten birthright with new life.

The expectations placed upon the new company were incredibly high. That which was left of the brand name A. Lange & Söhne and its identity bore the transfigured characteristics of a myth of an excellently hand-performed craft, timeless style, and quality without compromise. The competition against which the new Lange watches had to stand up against was thus not only the world of luxury wristwatches at the end of the twentieth century, but the company's own legacy.

However, the four wristwatches introduced in Dresden's castle delivered the proof that it was possible to simultaneously maintain and develop a legacy. The clever and emotional combination of both traditional and modern elements was successful in picking up the thread that had been lost forty long years previously. Günter Blümlein, whom I had the honor of personally knowing, understood how to answer the decisive question: How would a wristwatch look if Ferdinand Adolph Lange were to make one today?

That company founder Lange's virtues and value standards—contemporarily interpreted—are still en vogue today is proven by nothing if not by the unparalleled success of A. Lange & Söhne's watches in our reader's choice award, "Watch of the Year." Six wins and six showings among the top three in thirteen years speak an impressive language.

The city of Glashütte celebrates its 500-year anniversary in 2006, and it has all the reason in the world to revel: The seeds that Ferdinand Adolph Lange sowed more than 160 years ago have now borne fruit for the second time. During one human life there are few opportunities to experience the course of history up close. I have had the immense fortune to at least follow the second coming of the brand A. Lange & Söhne from close quarters.

Franz-Christoph Heel
Publisher ArmbandUhren

Contents

History & Histories

How it all began ...

The Master: Johann Christian Friedrich Gutkaes	6
An Exemplary Pupil: Ferdinand Adolph Lange	7
Project Glashütte	10
A Dynasty is Born	14
A. Lange & Söhne: On the Way to World Fame	16
A Lifetime of Progress	22
Chronological Historical Table	24

Walter Lange

When Time Came Home	26
Time and Again	28
Bibliography	31

A New Beginning after Reunification

Günter Blümlein: The Watch Maker	32
The "Delicate Little Flower Called Lange"	34
A Legend Becomes Time	42

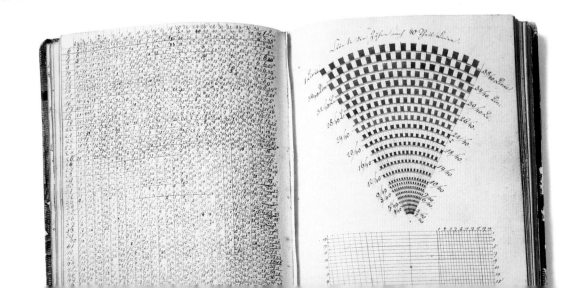

Contents

Collections and Classics

Lange 1	46
Lange 1 Moon Phase, Lange 1 Time Zone, Lange 1 Tourbillon, Grand Lange 1 Luna Mundi	
Saxonia	62
Arkade	66
1815	72
1815 UP and DOWN, 1815 Automatic, 1815 Moon Phase	
Cabaret	80
Cabaret Moon Phase	
Langematik	84
Langematik Perpetual, Anniversary Langematik	
Chronographs	92
Datograph, Lange Double Split, 1815 Chronograph, Datograph Perpetual	
Pour le Mérite	104
Tourbograph Pour le Mérite, Tourbillon Pour le Mérite	
Richard Lange	110

Manufacture and Makers

Portrait Fabian Krone	
The Electric Field between Tradition and Modernity	116
The Manufacture's People	
Heartfelt Handwork	120
Portrait Hartmut Knothe	
Right from the Beginning	138
A. Lange & Söhne's Success	
Around the World	142
The Collection of A. Lange & Söhne	148
A Word about the Third Edition / Masthead	156

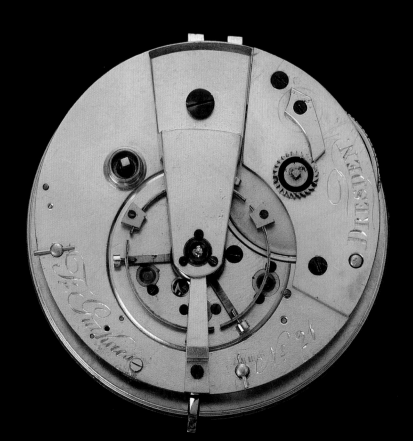

A rare pocket chronometer by J. C. F. Gutkaes with a regulator dial from about 1815.
(Photo: Dr. Crott Auctions)

How It All Began...

How It All Began ...

The Master:
Johann Christian Friedrich Gutkaes

Top: The arcades in the courtyard of Dresden's palace.

Top right: Johann Christian Friedrich Gutkaes (1785 – 1845)

Bottom: The five-minute clock constructed by Gutkaes in Dresden's Semper Opera house.

Augustus, the father of the Saxon people, did not only demonstrate his legendary strength (August was called "the strong one") on the battlefield—this electoral prince who reigned between 1694 and 1733 supposedly fathered over 200 children. Augustus had more interests than just women, however, loving other fine things in life such as architecture, painting, theater, music, and the sciences, a discipline still viewed as slightly godlike in those days.

During his regency, Augustus's royal residence in Dresden was lavishly renovated and expanded to resemble the style of the French royal seat in Versailles. Dresden was flourishing in the early eighteenth century and boasted many beautiful structures that were Venetian and Florentine in style such as the Zwinger and the Church of Our Lady. The beautified cultural metropolis was dubbed "the Florence of the Elbe," and the term "Saxon Baroque" was coined.

The collection found today in the Mathematics and Physics Institute of Dresden's Zwinger is rooted in the art gallery founded there in the sixteenth century. Over the course of several centuries, this collection of scientific machines—featuring, more than anything else, pathfinders and timekeepers originating in a myriad of cultures—has grown significantly. Saxon watchmakers to the royalty were high in the hierarchy and enjoyed a host of privileges. Johann Christian Friedrich Gutkaes (1785–1845), for one, lived in the castle tower as befit his office as the land's official timekeeper. The official time was displayed for all to see on the tower's clock. In 1841 Gutkaes, as the official watchmaker to the Royal Saxon Mathematics Institute, repaid the kindness shown to him by constructing a clock for the newly built Semper Opera House that could be seen from every seat in the audience. Due to the close quarters in the opera house, Gutkaes chose to place a digitally displayed five-minute clock above the stage, a masterpiece that garnered considerable attention in the western world.

In keeping with the tradition of the great clockmakers who had previously held his post, such as Johann Gottfried Köhler, Johann Heinrich Seyffert, and Johann Friedrich Schumann, Gutkaes paid close attention to further developments in clockmaking techniques, designing and selling high-grade pocket watches and presenting the up-and-coming land of Saxony with precision clocks to aid in establishing what was to become Central European Time at the end of the nineteenth century.

It must have been divine intervention that gave Ferdinand Adolph Lange, the foster son of an upright merchant family, access to this select trade, at barely fifteen years of age, by becoming an apprentice to the great Master Gutkaes.

An Exemplary Pupil: Ferdinand Adolph Lange

Ferdinand Adolph Lange was born on February 18, 1815, in Dresden, where his father, Johann Samuel Lange, a Silesian immigrant, was employed as a gunsmith for the royal guard. As to be expected of a blacksmith coarsened by a soldier's life during the Napoleonic Wars, the man was not the best of fathers. His wife left him, taking along with her their two children and giving the sensitive Adolph, her problem child, into the care of a merchant family in Dresden she was friendly with. It was clear that the intelligent and eager pupil would prosper with more schooling, and his foster family made many sacrifices to allow him to enroll in the city's Technical Trade School. There Lange got to know a young man of the same age named Friedrich Gutkaes, the royal watchmaker's son. While still in school, both young men began their apprenticeships with the older Gutkaes.

Top: Ferdinand Adolph Lange and Antonia Lange, née Gutkaes.

Top right: Johann Samuel Lange's proof of citizenship.

Right: A look into the living room of J. C. F. Gutkaes in the tower.

Right: Gutkaes and Lange's watch shop was located on the corner of Schloßstraße and Rosmaringasse in Dresden.

Far right: Gutkaes lived in Dresden's castle tower from 1842.

How It All Began ...

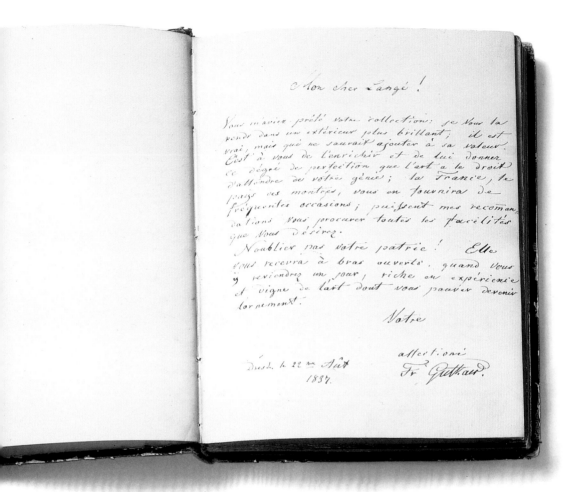

"Mon cher Lange": Lange's teacher Gutkaes wrote a personal recommendation into his journey notebook in 1837.

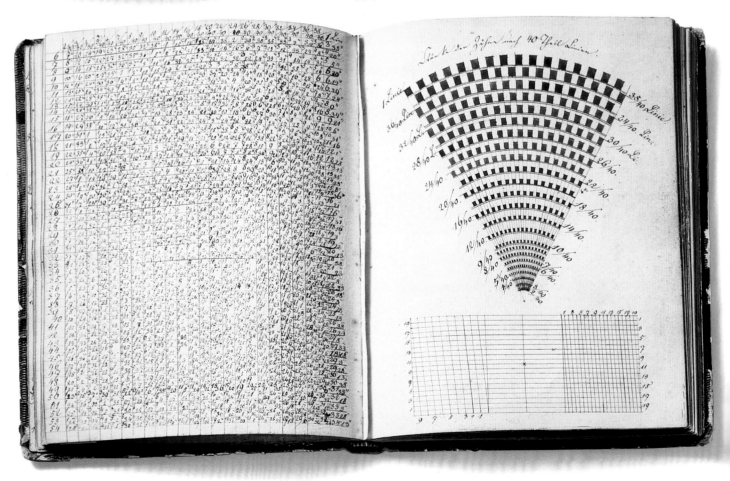

Bottom: Using microscopically small handwriting, Lange noted the calculations of the pitch diameters of wheels and graphically illustrated the sizes of the wheels across from them.

Johann Christian Friedrich Gutkaes was a very respected watchmaker and owned the finest and most reputable watch shop in Dresden. In addition, he fabricated fine watches and clocks himself.

The apprenticeships of Lange and Gutkaes lasted five years, from Easter of 1830 until Easter of 1835, due to the fact that they simultaneously attended the trade school, later to become the Polytechnical School. Master Gutkaes had meanwhile been appointed *mechanicus* of the Mathematics and Physics Institute and allowed his appreciative apprentices to look over his shoulder while he constructed precision clocks for the official observatory and public timekeeping service. Lange remained in Gutkaes's service for two extra years as a journeyman after his apprenticeship had officially ended, but the prospect of exploring the new European centers of watchmaking in London and Paris compelled him to begin his journeying in earnest in 1837.

Lange's Long Years of Journeying

The creative watchmaking that had found a home in the German-speaking parts of Europe (Nuremberg, Augsburg, Schaffhausen, and Strasbourg) during the Renaissance was giving way to a new, more privileged atmosphere at the courts of France and England. Continuously searching for the most precise timekeepers to be used for military and civilian seafaring, the monarchies supported watchmaking both financially and technologically.

Armed with a letter of recommendation from his master, Lange traveled by way of Mainz to Paris and to the famous chronometer maker Josef Thaddäus Winnerl, a pupil of the great Abraham-Louis Breguet, and the twenty-two-year-old Lange immediately felt right at home in this great center for watchmaking thanks to the French he learned at the Polytechnical School.

He remained with Winnerl for four years, quickly working his way up to the position of factory foreman while also studying astronomy and physics with Dominique François Arago to strengthen his scientific knowledge of the trade. Lange was not a big fan of trial and error, a concept that was highly acceptable in all other corners of the horological world. He filled his journeyman's notebook with numerous sketches and tables of mathematically based proportional relationships between gears and wheels according to the Parisians. The precision of the sketches leads one to believe that many of the era's exceptional watchmakers trusted Lange so much that they allowed him to study their watches and clocks extensively. He rendered in minute detail escapement systems, an adjustable Graham movement, pendulum compensation devices, repeater movements, and complicated chronometer escapements with constant force that he had found in the works of Moinet, Vuillamy, Wagner, Brocot, Lépine, and Breguet.

After remaining in Switzerland for many weeks on another leg of his journey, he found he no longer had the desire to visit England and also rejected Winnerl's offer of continuing to work with him in Paris. In 1837 Gutkaes had written in Lange's notebook, "Always remember your Fatherland! It will welcome you with open arms when you one day return rich in experience and having proven yourself worthy of the craft that can also enrich you." Around 1841 Lange knocked on his former master's door in Dresden, became the co-owner of Gutkaes's company, and sealed the relationship with a marriage to Gutkaes's daughter Antonia, thus binding himself even more to his homeland.

From Lange's sketch book: A chronometer escapement according to Arnold and a compensation pendulum according to Berthoud.

Project Glashütte

Saxony's Industry in the Nineteenth Century

The Industrial Revolution that had begun in England found fertile ground in Saxony at the beginning of the nineteenth century. Parallel to a flourishing manufacturing industry, numerous newly founded companies were being developed in industries such as cotton, where these companies had found a niche as suppliers for larger concerns according to the principle of division of labor. Due to its above-average economic and population growth, Saxony counted as one of the most successful industrial states in the world at the turn of the twentieth century.

Although the ideals of the French Revolution had failed on account of that country's pronounced feudal structure, a timid liberalization had arrived in other lands with cosmopolitan attitudes. Saxony's advanced, machine-powered agricultural industry was able to feed the land's ever-increasing population well, and potential agricultural workers had no problem finding new jobs in growing industrial companies nestled in the Erzgebirge mountain range and at its foot, in an area called the Vogtland. As it was still necessary to purchase tools and production equipment from foreign countries (mainly England, France, and Belgium), Napoleon's Continental System, a trade embargo from 1806 to 1812 designed to isolate and destroy England's economy, hit Saxony's industrial concerns especially hard. Just as it had been decided to invest heavily in Saxon industrial companies, Napoleon lost his Russian campaign and the embargo ended. Ironware and textiles from Britain that had been held back for years flooded the market in Saxony and continental Europe. Both small manufactories and newly founded industrial companies located in the mountains of the Erzgebirge were not (yet) able to withstand the price pressure. These companies' employers, still feudally structured, retracted their capital, and Saxony's workforce was thrust into poverty.

Adolph Lange experienced these times of crisis during his youth. It was only on an initiative of the new secretary of the interior, Bernhard Graf von Lindenau (King Frederick Augustus I had died after fifty-eight years of regency), that Lange was even able to attend the Polytechnical School. Von Lindenau's concept included supporting schools such as these in order to forge an indigenous mechanical engineering industry to reduce dependence on foreign powers.

Lange's Basic Concept: Independence

As a student at the technical trade school, Lange had avidly listened to the ideas of his instructor, Wilhelm Gotthelf Lohrmann, concerning the land's industrialization. During his journeyman years in Paris, the Saxon railroad had commenced service, making its first connection between Dresden and Berlin in 1839, and precision clocks and pocket chronometers were in great demand to aid in distributing and keeping Saxon standard time.

After helping the Gutkaes & Lange company to a remarkable economic upswing with some of his own finely constructed complicated pieces and astronomic pendulum clocks, the young man took an important step that would later change the course of his life. In a letter written to Privy Councilor von Weissenbach, he sketched out his idea for the foundation of a pocket watch factory in the "structurally poor region" of the Erzgebirge mountain range.

The Product

Lange emphasized the fact that he could not approve of a new watch factory in Dresden due to the high cost of living there, and he referred to a failed project initiated by the French government that had taken place while he was in Paris of choosing Versailles as the location for a factory to produce watches for the people. The suggestion of bringing work, pay, and bread to the Erzgebirge area, a region in danger of total impoverishment, by way of a watch factory caught the ministry of the

Top: Lange's first workshop was located in a rented building in the city.

Right: In 1873 Lange moved into the newly constructed building at Altenberger Str. 1, today referred to as the historic family domain.

interior's attention, especially since Lange was able to formulate clear ideas concerning both the product and the factory and had minutely calculated the amount of state financial support needed.

His plan concentrated on one single watch model at the beginning, the construction of which he had perfected and simplified as far as possible using mathematical calculations and whose components could be efficiently manufactured on partially self-constructed machines (i.e., lathes operated by using a fly wheel instead of a crank, new engine-driven milling machines). The crux of his movement construction was the Glashütte pallet escapement movement, later to become legendary, and the large-surfaced three-quarter plate. Lange described his new watch design in a detailed plan: "I am combining the pleasing shape of the Swiss cylinder

Three rare, fine pocket watches of best Lange quality. Above left a hunter with an ornamental case according to Prof. Graff (around 1897); above right a hunter in Louis XV style (around 1885), and below a half hunter from around 1876 (photo: Dr. Crott Auctions).

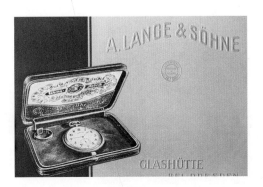

Project Glashütte

Left: Around 1900 the historic family domain had already been added on to several times.

Right: After being bought back and renovated, the regal building literally glowed. The new address has been Ferdinand-Adolph-Lange-Platz 1 since December 7, 2001.

watch with the large power reserve and the acknowledged precision of the very expensive and uncomfortable English pallet watch. Thus improved by me, this timepiece will guarantee the best rate precision, will be lighter to wear than the heavy cylinder watch, will comprise only two-thirds as many parts, and will not be outfitted with light, fragile components."

Division of Labor

The detailed list of the serially performed processes that Lange had already assigned to certain people accompanying his letter acknowledges the fact that Lange had put some serious thought into the principle of division of labor. His thoughts spanned all the way to the makeup of the core workforce that he wanted to recruit "if possible from one village, or if that is not possible, then from those that are closest to each other" in order to retain a feeling of team spirit even on the way home after work. The fifteen apprentices shouldn't be too young, possibly with some experience and talent in woodcarving or weaving, and after finishing their apprenticeships they should be able to manufacture twelve watches per week, or 600 per year. Lange put a plumber to hammer brass on his wish list and also asked the privy councilor to think about offering a scholarship to a trained metalworker so that he could be sent to Switzerland as an apprentice to a case maker. In this way it would be possible to emancipate Saxony from foreign companies in this area of manufacture as well.

The Training Concept

Lange needed to resend his concept in January of 1844 (after six months without a reply to his suggestion, he had even considered manufacturing marine chronometers for the German, Dutch, and Belgian navies), but then things began to happen very quickly. A commission picked the little city of Glashütte from the various communities of Zwickau, Bautzen, and Dresden as the ideal place for Lange's watch fabrication, and the Royal Saxon Ministry of the Interior gave him a loan of 5,580 talers as well as the nonrepayable sum of 1,120 talers for the purchase of the tools and machines needed.

The commission also followed Lange's suggestions to the letter on another point: the apprentices, freed from all financial responsibilities during their three-year apprenticeships, would pay their premiums back to Lange at the end of their training period in small installments. Lange himself was obligated to pay the sum of 5,580 talers back to the Ministry of the Interior in seven installments. All that the fifteen apprentices had to do in return for the exceptionally generous loan granted during trying economic times was to work at Lange's factory for a period of at least five years after their apprenticeships were over for relatively low wages.

In the year 1845 Lange moved to Glashütte despite the fact that his original plan had called for the education of the fifteen apprentices to be completed in Dresden near the Polytechnical School. He rented a building

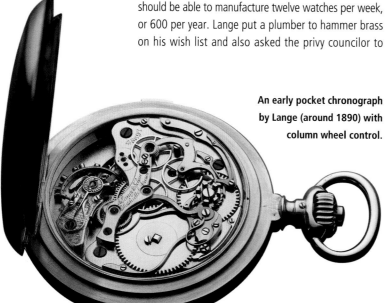

An early pocket chronograph by Lange (around 1890) with column wheel control.

and an apartment and began to set up a workshop for the apprentices. Initially he was aided by former pupil Adolf Schneider, watchmaker Louis Müller, and—at least part of the time—brother-in-law Bernhard Gutkaes. Lange personally chose his future apprentices from the area surrounding Glashütte, and on December 7, 1845, the factory building was finally officially opened, and the apprentices were welcomed to the school. This was an historic date.

The Region

A. Lange & Comp. taking up work represented the beginning of a sensational experiment in social and economic policy, one whose effects can still be clearly felt today in the region surrounding Glashütte.

According to his division of labor principle, Lange assigned the apprentices to a certain area of production or engineering after completing their training periods. During the obligatory five years of working for Lange, the journeymen increasingly specialized in their learned skills, for in Lange's original plan it was stated that he wanted to inspire the apprentices who had "become serious and manly" workers during their apprenticeships to independence.

The first of Lange's workers became self-employed in 1848 and founded several smaller ateliers for specialized supply parts within the city of Glashütte. These mechanical workshops made things easier for the main company by prefabricating components such as flat parts and bridges, cutting out wheels, milling gears, turning spring barrels, drilling jewels, manufacturing hands, and gold-plating various movement parts. In Lange's own workshop, these components were all finely finished and assembled to form watches. In order to encase the regulated movements, Lange founded—more or less unwillingly—his own case workshop and acquired a guilloché engine to decorate the casings appropriately. He paid for such unexpected expenditures from his and his wife's own private funds and even took on personal debt.

It was only Ferdinand Adolph Lange's selfless enterprises that allowed the company's initial difficulties (in reality, production was not even close to the calculated twelve watches per week) to be overcome. However, it was the continual improvement of the Lange pocket watches' quality and technology as well as the cautious expansion of the model palette that created a solid basis for good sales in his brother-in-law's retail shop in Dresden's best district.

A heavy gold hunter's watch with minute repeater from around 1890.
(Photo: Dr. Crott Auctions)

A Dynasty is Born

Friedrich August Adolf Schneider (born in 1824) came to Glashütte with Adolph Lange to aid his former instructor in setting up the company's parent factory. In the meantime he had married the younger sister of Lange's wife Antonia Gutkaes and thus belonged to the family. In 1851 he was one of the first to set up a second base of operations alongside his work at Lange's factory: By 1855 his company became known as Glashütter Uhrenfabrik Adolf Schneider. Schneider was able to fall back on the Glashütte infrastructure, well set up by then, and—with the blessing of his brother-in-law—profit from the newest acquisitions in the continuously developing technology of movements.

Chronometer maker Julius Assmann (born in 1827) came to Glashütte in 1850 to enter Lange's service, filling the position of an experienced foreman that had become vacant because of Schneider's gradual withdrawal from the factory. A short two years later, Lange inspired the Stettin native to found his own watch factory in order to make full use of the growing amount of suppliers in and around Glashütte. Like Schneider before him, Assmann also profited from the newest in technological manufacturing and movement acquisitions, to which all three watch manufacturers almost simultaneously had access. This cooperation made up of fair competition worked even better when Assmann married Lange's oldest daughter Marie (his second marriage) in 1865 and thus entered the elite circle of the Glashütte watchmaker dynasty.

Although he didn't belong to the Gutkaes/Lange family clan, it should be mentioned at this point that Carl Moritz Grossmann was seen as an important member of the original Glashütte manufacturer's scene. Grossmann (born in 1826), Lange's former pupil and at the time co-owner of Gutkaes's store in Dresden, settled in Glashütte as a manufacturer of watches and measuring instruments after a long absence from the trade spent in diverse armed forces. He specialized in precision pendulum clocks and marine chronometers, sliding gauges and micrometers, constateurs, short-term timers, and revolution counters. For this reason Grossmann did not use Glashütte suppliers for the manufacture of his own watches, but rather supplied the other factories with his measuring instruments and worked out a number of improvements for machines, pocket watches, and pendulum clocks together with Lange.

Since Grossmann was also able to make his mark with a number of sensational written standard works on the design of watch movements and movement parts, he was no doubt the right choice for director of the German School of Watchmaking, whose founding in the year 1878 can mostly be attributed to Grossmann's initiative and expertise. The German School of Watchmaking was generously supported by all four Glashütte watch manufacturers, who supplied it with instructors, teaching and class material, and a know-how garnered in thirty years of pioneer work in industrial watch design. The school immediately developed into the nucleus of a new generation of watchmakers.

Death is Not the End

When Ferdinand Adolph Lange died on December 3, 1875, his noble life's work was still greatly in evidence. In thirty years of self-sacrifice, he had realized his magnanimous plan of bringing jobs and affluence to the region surrounding Glashütte, against the odds and despite small setbacks. The population of Glashütte rose from 1,030 to 1,722 inhabitants within those thirty years: where one generation's previous hard work with bare soil had to suffice to eke out a living, the watch industry and its related branches were already feeding one-fifth of the valley's population in that time. Lange remains today—150 years after the tender beginnings of industrialization in the Müglitz Valley—famous for being the bringer of light to a community that he also led as mayor for eighteen years. It was his initiative that advantageously changed the face of this rural community. Lange had goose ponds and manure piles removed, streets and bridges built, and the Prießnitz river bed reinforced with sea walls. His "Lange foundation" developed into both accident insurance and pension money for disabled and aging watchmakers, and Lange and his company ended up creating new living space for sixty families with loans and construction projects in Glashütte and the surrounding area.

Julius Assmann

To honor the fiftieth anniversary of the watch industry, the city of Glashütte erected a monument to its founder on August 31, 1895.

A Dynasty is Born

Unmistakably Glashütte watchmaking: Above left a gold hunter's watch constructed by Richard Lange (around 1905); above right a gold pocket watch by Adolf Schneider (around 1890); in the center an observation chronometer constructed by Lange's former employee Fridolin Stübner (around 1900); below left "homework" from the German School of Watchmaking (Walther Urban); and below right a rare gold hunter's watch constructed by Moritz Grossmann around 1870. (Photo: Dr. Crott Auctions)

On the occasion of the fiftieth anniversary of the watch industry in Glashütte, the city had a monument built for its most honorable citizen in the shape of an obelisk made of dark green syenite with his portrait on a bronze mezzo relief.

Despite all of this, it remains a fact that while Lange was in Glashütte he only laid the foundation. The great, worldwide success of the Glashütte watch industry in general and of the brand A. Lange & Söhne in particular was actually achieved by the second and third generations.

A. Lange & Söhne: On the Way to World Fame

When Lange's son Richard (born in 1845) joined the company in 1868, the name was changed to A. Lange & Söhne. Richard had inherited his father's great talent for technical perfection and brought a well-founded knowledge of all the disciplines of watchmaking into the company after a four-year journey with stops with such masters as Paris chronometer maker and Winnerl-pupil Simon Vissière, and London's Friedrich Böhme. He obtained numerous patents (among others for a quarter-hour repeater, a jumping second, the Nivarox balance spring, and a power reserve display) and extended production to the rooms of the new main factory building in 1873.

In 1871 Richard's younger brother Emil (born in 1849) joined the company. At the beginning, he occupied himself completely with the firm's business management and made contacts in Switzerland and elsewhere in order to obtain dials, springs, balance springs, and later complicated mechanical components that could not be manufactured in Glashütte in the amounts needed.

Otto von Bismarck's foundation of the German Reich as a confederation of states after the German-French war brought about a strong nationalistic feeling that was enhanced when Prussia's King William I was proclaimed German emperor. The following forty years until the outbreak of World War I were the most successful and dazzling for the German watch industry and witnessed the development of A. Lange & Söhne into an exclusive luxury brand.

Alongside beautiful pocket watches in lavishly decorated gold cases, Lange produced more and more complicated watches with tourbillons, repeater mechanisms, chronographs, and perpetual calendars around the turn of the century. A second product line chiefly comprised pocket chronometers with lever escapements that were tested for their rate precision by the royal observatory in

Emil Lange

A glance into Emil Lange's private study (Limmer watercolor painting, around 1897).

On the Way to World Fame

Top: Magnificent gold hunter's watch from around 1898. Emperor William II gave this Lange watch to the Turkish sultan Abdul Hamid II on the occasion of his first visit to the Ottoman Empire.

Right: One-minute tourbillon from around 1900. The miniature enamel painting of Minerva at Dresden symbolizes the victory of world peace via artistic craftsmanship, technology, and medicine. It was introduced by Emil Lange at the World's Fair in Paris in 1900.

Leipzig. Around the year 1898 the production of marine chronometers and deck and observation watches for the imperial navy was added.

Although the company had only moved into the factory building at Altenberger Str. 1 in 1873, continuously increasing production made an annex onto the main factory necessary in 1898. Just eight years later, in 1906, a new storey was already added onto this building.

In all of these years, not a whole lot had changed in the manufacturing process of the high-quality Lange watches, and various Glashütte companies prepared to fill the vacuum in the middle and lower-end price classes—among others, companies like the Glashütter Präzisionsuhrenfabrik Akt. Ges. and the Nomos-Uhr-Gesellschaft, operating with machine-produced mechanisms and even movements purchased in Switzerland. The only quality concession that managing director Emil Lange (after his brother Richard left the company, he managed it alone) was prepared to make restricted itself to the introduction of the brand Deutsche Uhrenfabrikation (DUF, distributed by wholesaler Dürrstein in Dresden) whose movements had to suffice without screw-mounted chatons, gold screws in the balance, and regulation in only three positions (top-quality Lange watches were regulated in five positions).

With the outbreak of World War I, the market for the expensive Lange watches of both classes collapsed, and the company was able to more or less hold its head above water by manufacturing marine chronometers and deck and observation watches. Since the inexpensive everyday watches made by the other Glashütte brands were based on the import of movements and parts from Switzerland, these manufacturers went bankrupt in the first year of the ban on imports. After the end of the war, an initiative of the central association of German watchmakers led to the founding of Deutsche Präzisions-Uhrenfabrik Glashütte in Sachsen eGmbH (DPUG) that was to produce reasonably priced large-series pocket watches on modern machines. Emil Lange's son Otto reacted by designing a simplified, machine-manufactured watch movement caliber in the (Swiss) bridge method of construction that, however, visually suggested the famous three-quarter plate.

Emil Lange was not prepared to accept this new method of production toward the end of his managing days and handed the leadership of the company over to his sons Otto, Rudolf, and Gerhard in May of 1919.

On the Way to World Fame

Gold pocket watch with up and down movement and rattle winding that can be changed to crown winding from around 1917. Gold-plated movement in top quality with screw-mounted gold chatons.

Hard Times for the Third Generation

The reasonably priced watch for everyday use called to life by Otto Lange was based on the OLIW caliber featuring a 43 mm diameter. The Original Lange International Movement (Werk) represented a movement assembled for the first time with machine-manufactured parts. The preparations (especially measuring and sketching the components) took several years, and the promising OLIW caliber only went into production at the beginning of 1923.

The advent of the world's economic crisis, however, nipped the company's upward turn in the bud: Two-thirds of the workforce had to be let go, and the rest of the employees worked short shifts. Price calculations became virtually impossible—at the beginning price lists were re-done every two weeks, although by the middle of the year they were done every day. The price for one OLIW watch exploded within a year to the tune of several hundred million Reich marks.

The lifting of the Swiss-timepiece import embargo at the end of 1924 caused a rude awakening. During the previous ten years the Swiss industry, untouched by war, had been able to create an enormous palette of reasonably priced timepieces for daily use, which also contained fashionable wristwatches, leaving the solid but old-fashioned and wickedly expensive Glashütte production hardly a chance to endure. After Assmann and DPUG went bankrupt, all chances for prevailing were once again placed on the shoulders of Lange and the newly founded Uhrenfabrik OLIW oHG in Glashütte, a factory that, together with the registered brand A. Lange

From 1913 Lange had access to a standard timekeeper controlled by a long-wave time signal to synchronize all of the company's chronometers.

Right: Two rare Lange wristwatches. The cushion-shaped watch is from the World War I era and is outfitted with a DUF three-quarter plate movement. The tonneau-shaped watch was produced around 1935 and is outfitted with a Swiss bridge movement purchased from Audemars Piguet. (Photo: Dr. Crott Auctions)

On the Way to World Fame

Marine chronometer especially equipped for use as a chronometrier instrument in military field technology containing electronic second contact, balance-stop mechanism, and external mechanism to set the hands. From around 1941. (Photo: Dr. Crott Auctions)

& Söhne, secured the survival of the company. Alongside Lange the only producers left in town included the successor organization to DPUG called Uhren-Rohwerke-Fabrik Glashütte (Urofa), owned by the Girozentrale Saxony bank, and Glashütter Uhrenfabrik (Ufag, which owned the better-quality Tutima brand).

During the 1930s, the new wristwatch trend pretty much went unnoticed by Glashütte. Lange produced a small collection of round and rectangular wristwatches for men and women, but all of them were outfitted with Altus ébauches purchased in Switzerland.

The company experienced an economic upturn after the National Socialist Party came to power and contracted for marine chronometers and observation watches for the military. In a very short time, Caliber 48 (featuring a 48 mm movement diameter and subsidiary seconds; also available as Caliber 48.1 with sweep seconds) was created from the flat OLIW pocket watch Cal-

On the Way to World Fame

Top: Fighter pilot observation watch from around 1941. Unique silver version made for Hermann Goering. (Photo: Dr. Crott Auctions)

Right: Exceptionally rare Luftwaffe B Watch with graduated gauge scale on dial, produced by the DUF (Deutsche Uhrenfabrikation Glashütte) based on Lange's Caliber 48. (Photo: Dr. Crott Auctions)

iber 90, serving its time in various military watches. Due to this caliber's high demand, it was also assembled by other companies.

As a former military supplier located in the Russian Zone after the four-power agreement had taken place, Lange had a very hard time after the war. As bombs destroyed the company's main production building (called the Werft, or hangar) on the last day of the war (May 8, 1945), there were no longer any production machines left to dismantle at Lange. Instead the occupying forces took Lange's intellectual property: all of the technical documents concerning Caliber 48 and the marine chronometer had to be handed over. The Lange marine chronometer is still manufactured today in Russia according to these plans. The B Watch and the B Chronometer (outfitted with Caliber 48) were manufactured from remnant stock. In addition, the OLIW pocket watch was prepared for production, and a wristwatch movement (Caliber 28) was developed.

When it was expropriated on April 20, 1948, the independence of the company A. Lange & Söhne ended, including the company's own determination of quality and watch collections produced under the brand name Lange VEB (Volkseigener Betrieb or Company of the People) until 1951. A combining of the Glashütte companies A. Lange & Söhne, Urofa/ Ufag, Otto Estler, Lindig & Wolf (Liwos), R. Mühle, and Gössel & Co. took place on July 1, 1951, and Glashütter Uhrenbetriebe (GUB) was created.

On this day, all of these historical company and brand names were erased.

A Lifetime of Progress

Ferdinand Adolph Lange was not only the founder of the so-called Glashütte school of watchmaking with its distinctive characteristics and detailed solutions in watches and their design. Lange—strictly speaking one should make reference to the entire Lange dynasty, for his sons and grandsons were just as involved—transmitted valuable horological impulses far outside the borders of Saxony, which were received in Switzerland as well as elsewhere with great interest.

F. A. Lange had studied the technology of tools during his journeyman years and used this knowledge as the basis for research on his ambitious industrial project far away from established watchmaking locations. More than thirty years before the introduction of the meter as the official length of measurement in Germany, Lange was using the metric system to calculate the dimensions of his mechanical mechanisms, and had created meticulous, extensive conversion tables for the measuring units of foot, inch, and line still used respectively in England, France, and Switzerland. This straightforward technocrat did not understand why one would want to fool around with feet, inches, lines, and hundredths of lines in the mathematically non-user-friendly base-twelve system when the Paris Meter Convention had established a length of measurement based on the decimal system in 1795. Lange's first workshop in Glashütte included some measurement tools that he had constructed himself such as the jar micrometer, a lever dixième gauge, and a sliding gauge with vernier adjustment.

Until that point, only arbitrary accounts based on trial and error had been noted in watchmaking to cut wheels and gears. Contrary to the norm, Lange developed cyclical toothed shapes according to his own very detailed drawings that rubbed against each other with only a minimal amount of friction. He also created very extensive proportional tables with which the pitch diameter, the intensity, and the tooth base could be ascertained in advance to calculate the diameter of any given toothed wheel, including those with an odd number of teeth.

And where materials were concerned, Lange was also ahead of his time: He taught his students to mill gears right out of steel plates, instead of doing it the Swiss way by filing, and he developed new milling cutter shapes especially for this. Smaller and larger oscillation wheel lathes replaced the still widespread watchmaker "violin bow-driven" lathes with which the chasing tool had to be lifted and put in the right place with every turn of the piece. And, last but not least, Lange placed a great amount of value on delivering completely regulated watches to jewelers and watchmakers. Until 1845 it was an industry norm for the watchmaker who sold the watch to spend a few hours regulating or even checking the spring of the untested watch movements.

Continuous Development

F. A. Lange's first apprentices continued to develop his special techniques after they had become independent. Screws manufactured in Glashütte, for example, weren't rolled, as was the usual process in Switzerland, but were rather cut according to a method created by Lange. Also profiting from continuing research in Glashütte were the following developments: the lever movement typical of the region with its hidden pallet stone and variously shaped surfaces on the entry and exit sides, the new Glashütte compensation balance, and the balance's regulating screws fastened in screw threads with a slit for securing the screw by suspension.

With the shift from key winding to crown winding, Adolph Lange designed a two-piece winding shaft, whereby the intervention of the winding gear in the crown wheel was not entirely dependent on the case fitting. Later a spring system was developed to control intervention energies and to guarantee improved hand-setting mechanics.

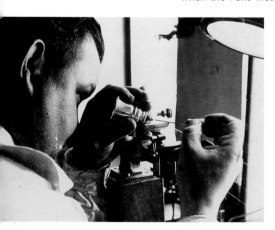

Top: Inspecting a jewel after finishing.

Right: Setting the balance spring.

The development of the precision pocket watch under F. A. Lange went in three directions in the 1860s and these deserve special mention: constant force; the large, jumping sweep second hand; and the resettable large second hand (chronograph), all of which were based on the simpler and more reliable lever escapement and not the delicate chronometer escapement. Adolph Lange received patents for crown winding, a device to set hands, and a quarter-hour repeater, among other things.

Like Father...

Richard Lange lovingly carried on with his father's life's work, continuing to improve the chronograph and the "jumping" second hand, but also leading the way to the future by patenting an up and down movement (power reserve indication). As Reinhard Meis's book *A. Lange & Söhne—Eine Uhrmacherdynastie aus Dresden* [A. Lange & Söhne—A Watchmaking Dynasty from Dresden] illustrates for the first time, Richard Lange may also be seen as the originator of the Nivarox balance spring since he discussed the possibility of adding beryllium to the spring to increase its hardness in his 1930 patent entitled "Metal Alloys for Watch Springs." The Swiss national Reinhard Straumann, who coined the term Nivarox (derived from "nicht variabel und oxidationsbeständig"—not variable and able to withstand oxidation), had to base his 1931 patent on the Lange patent. Straumann is attributed the technical realization of the Nivarox spring by supplementing alloy additives, but it was without a doubt Richard Lange who paved the way for the beryllium-nickel alloy still used today. And by the same token, it is certainly of interest to mention that A. Lange & Söhne is once again manufacturing its own balance springs for the first time in the twenty-first century—springs that are at least in part based on Richard Lange and his employees' know-how.

The list of patents and registered designs that can be attributed to Lange's company and its respective owners stretches from case design to escapement systems (chronometer, *Karussell*), individual mechanisms and devices, and material combinations. The improvements for details were countless, spanning every aspect of the manufacture and design of fine watches, including lubrication, and have flowed into all of the products issued by the house of Lange over the decades. Because of careful record-keeping today in Glashütte, these are things that can once again be proudly recalled.

Richard Lange

Top: A glance into the watchmakers' workshop, around 1920.

Chronological Historical Table

1815 Ferdinand Adolph Lange is born on February 18 in Dresden.

1830 Lange begins his watchmaking apprenticeship with Johann Christian Friedrich Gutkaes.

1835 Lange completes his watchmaker apprenticeship with honors.

1837 In August Lange finds his way to Joseph Thaddäus Winnerl in Paris via Mainz.

1841 Gutkaes makes the five-minute clock for the Semper Opera House in Dresden. Lange comes home from his wandering years and begins working with his former instructor Gutkaes. Adolf Schneider (b 1824) begins his watchmaker apprenticeship with Gutkaes.

1842 Lange gets his master watchmaker title and marries Gutkaes's daughter Antonia. Gutkaes is named watchmaker to royalty. Carl Moritz Grossmann begins his watchmaker apprenticeship with Gutkaes.

1843 On June 25, Lange presents his industrialization project to the privy councilor of Weissenbach.

1844 On January 11, Lange sends the privy councilor a written reminder of his project. On May 3, he makes his final request to the Saxon government.

1845 On May 3 the government commission's choice of place for Lange's industrialization project falls on Glashütte. On May 21 Lange and the Royal Saxon Ministry of the Interior sign a contract. On December 7 the watch company Lange & Comp. is founded in Glashütte. Lange's son, Richard, is born on December 17.

1847 Schneider marries Gutkaes's youngest daughter, Emma. Grossmann becomes an employee of Lange.

1848 Founding of Lange's own case workshop. Journeyman Jungnickel becomes an independent wheel miller. Journeyman Weichholdt becomes an independent manufacturer of escape wheels, later also making gears. Journeyman Lissner begins manufacturing screws. Journeyman Gläser becomes an independent manufacturer of hands. F. A. Lange becomes mayor of Glashütte, remaining so for eighteen years.

1849 On August 30 Emil Lange is born.

1850 Julius Assmann is hired as a watchmaker at Lange.

1851 Schneider founds his own watch company in Glashütte.

1852 Assmann becomes an independent watch manufacturer in Glashütte.

1854 Grossmann founds his own watch company in Glashütte.

1861 Lange makes a pocket watch with jumping seconds (seconde morte) and only one spring barrel.

1863 Lange makes a pocket watch with a mechanism to stop time (chronograph).

1864 Introduction of the characteristic three-quarter plate so that the gear train has a stable and stress-free position.

1865 Assmann marries F. A. Lange's eldest daughter (second marriage).

1866 Lange has crown winding and a simple quarter-hour repeater patented in the United States. Lange develops a calendar watch with date, day, month, and moon phase.

1868 Richard Lange enters his father's business as co-owner. The company's name is changed to "A. Lange & Söhne"

1869 Lange receives a U.S. patent for an improved hand-setting mechanism.

1871 Emil Lange enters his father's business as co-owner.

1872 Richard Lange marries the daughter of Berlin clock manufacturer Roessner.

1873 Lange's company moves into its new main factory building at Altenberger Straße 1.

1874 Lange presents the smallest women's lever movement with display of seconds and a 25 mm movement diameter.

1875 Ferdinand Adolph Lange dies on December 3.

1877 Lange receives a German Reich patent for its "jumping seconds."

1878 Founding of the German School of Watchmaking in Glashütte; lessons begin on May 1 under the direction of Grossmann.

1879 Lange receives a German Reich patent for its UP and DOWN movement.

1882 Production of a pocket chronometer with fusee and chain transmission.

1885 A. Lange & Söhne employs sixty people.

1887 Richard Lange leaves the company for health reasons. Brother Emil remains as sole director.

1888 Richard Lange receives a German Reich patent for improved "jumping seconds."

1891 Richard Lange receives two German Reich patents for improved chronometer escapements.

1892 Lange registers a German Reich utility model for a pocket watch outfitted with a minute counter.

1896 The Lange watch factory gets electricity with the aid of a water turbine.

1898 Renovation of the Lange main factory building to add new production rooms. Beginning of serial production of marine chronometers and observation watches.

1899	Lange registers a German Reich utility model for a "pocket watch movement with rotating parts" (Karussell).
1906	Further construction on the factory (back building gets a new floor).
1908	New, larger power station starts service.
1919	Emil Lange hands the company reins over to his sons Otto (b 1878), Rudolf (b 1884), and Gerhard (b 1892).
1920	Otto Lange registers a German Reich utility model for the OLIW watch.
1924	The import ban on Swiss watches, initiated in 1914, is lifted. On July 29 Walter Lange is born.
1925	Founding of the OLIW oHG watch factory in Glashütte.
1930	Richard Lange receives a German Reich patent for a beryllium-nickel alloy.
1938	Production begins on the Caliber 48 family for pilot's watches and navy observation watches.
1945	On May 8, the last day of the war, the factory is bombed. Production of the Caliber 48 is restarted using remnant parts.
1946	On March 18 the company is impounded by the Soviet military administration; on April 26 it is given back to its owners.
1948	On April 20 the company is once again impounded and expropriated. The name is changed to Mechanik Lange & Söhne VEB (Volkseigener Betrieb—Company of the People).
1951	On July 1 Glashütter Uhrenbetriebe (GUB) is founded by combining the existing companies A. Lange & Söhne, Urofa/Ufag, Otto Estler, Lindig & Wolf (Liwos), R. Mühle, and Gössel & Co.
1990	On December 7, exactly 145 years after the original founding of the company by his great-grandfather, Walter Lange registers Lange Uhren GmbH in Glashütte and globally registers the brand A. Lange & Söhne.
1994	In October the first Lange watches of the modern era leave the production rooms of the former precision pendulum clock factory Strasser & Rohde, the new company headquarters of Lange Uhren GmbH. The new models are Lange 1, Arkade, Saxonia, and the Tourbillon Pour le Mérite.
1995	Walter Lange gets the key to the city of Glashütte. The Lange 1 model is voted Watch of the Year by the readers of the magazine *ArmbandUhren*.
1997	Lange 1 is once again voted Watch of the Year by the readers of *ArmbandUhren* and the national newspaper Welt am Sonntag.
1998	Opening of the second production workshop (Lange II) in the former Archimedes adding machine factory, directly next door to the first factory building. Langematik is voted watch of the year by the readers of *ArmbandUhren* and *Welt am Sonntag*.
2000	Together with its LMH sister companies Jaeger-LeCoultre and IWC, Lange Uhren GmbH is sold to the Richemont Luxury Group at the end of July.
2001	On October 1 Günter Blümlein dies after a short, difficult illness. On December 7 the renovated Lange main factory building once again houses production workshops, a watchmaking school for apprentices, and management offices. The square in front of Glashütte's train station is renamed Ferdinand-Adolph-Lange-Platz.
2002	The Langematik Perpetual model is voted Watch of the Year by the readers of *ArmbandUhren* and *Welt am Sonntag*.
2003	On the property between the production workshops Lange I and Lange II, a generous new glass building known as the Technology and Development Center is erected to house various new departments, most especially that of the balance spring production.
2005	On July 8 the first thirty-six sets of the three Lange 1 Time Zone variations are delivered to thirty-six Lange dealers all over the world by thirty-six watchmakers from the manufacture and televised in real time via satellite on large video screens. Lange Uhren GmbH buys the former brewery's building and property on Altenberger Straße.
2006	The Lange 1 Time Zone is voted Watch of the Year by the readers of *ArmbandUhren* and *Welt am Sonntag*. At the SIHH in Geneva, the Datograph Perpetual and the three-handed Richard Lange timepieces are officially launched. The city of Glashütte celebrates its 500th anniversary.

When Time Came Home

Walter Lange doing what he loves best: studying a vintage Lange pocket watch. His opinion is definitive when new movements are created in typical Lange style.

Walter Lange was born on July 29, 1924, the fourth child of Rudolf Lange (1884–1954) in Dresden. Before he finished his watchmaker training, the not-quite twenty-year-old Lange was drafted and badly injured near East Prussia's Königsberg (Kaliningrad) in February 1945. After an adventurous odyssey across the Baltic Sea and Scandinavia, he arrived home in Glashütte on sick leave shortly before the war ended, just in time to witness Russian bombs destroying Lange's factories and the invasion of the Red Army.

At the time, the Lange company was jointly owned by Walter's father Rudolf and his uncles Otto and Gerhard. When the firm was impounded by the Soviet military administration in March 1946, the family was able to successfully register a protest. But three days after the sequester order went into effect, on April 20, 1948, the company was expropriated and turned into a "people's company" (VEB or Volkseigener Betrieb). The former owners were barred from even entering the production buildings.

Since Walter Lange was not yet one of the owners, he was able to continue working at the company, although he soon fled to the West after he refused to become a member of the FDGB (the so-called free union), an act the government retaliated against by sending him to work in the uranium mines.

When the end of the GDR era finally became fact, and the reunion of the two Germanys was becoming tangible, Walter Lange took heart and reawakened the dream that he had never quite given up of actually reinstating "his" brand. At the International Watch Co. in Schaffhausen he found in Günter Blümlein an enthusiastic and energetic ally who supplied the financial, technical, and industrial support of the prevailing structures within the LMH watch group (Les Manufactures Horlogères, majority of capital held by VDO/Mannesmann).

Both "face" and design traits of the new Lange wristwatches were the product of a highly motivated team, although the final okay of the sketches and prototypes produced by movement designer Helmut Geyer's department and a design team headed up by Anthony de Haas in recent times is always given by Walter Lange, the "keeper of Glashütte's watchmaking tradition."

Even between creation cycles for new models and model variations, the now eighty-two-year-old great-grandson of Ferdinand Adolph Lange hardly has time to

Walter Lange

Three generations of Lange: Walter Lange at the family gravesite. The photo at right shows him speaking with German President Johannes Rau at the celebration for the ten-year anniversary of the firm's refounding on December 7, 2000.

Walter Lange is a proud recipient of the Saxon Order of Merit.

rest. As the founder's inheritor and a living witness to former times, Walter Lange takes his role as the ambassador of his brand very seriously. Duties at receptions and new dealership openings are some of the important functions that keep him busy for the entire year.

In 1995 Walter Lange received the key to the city of Glashütte, an honor bestowed for the second time in history to this Dresden family that has done so much for this little town and its surrounding area.

Saxony's former minister-president, Kurt Biedenkopf, honored the untiring champion of German unity and the rebuilding of eastern Germany with the Saxon order of merit for his life's work: Characterized by highs and lows, the dream he kept alive through the years proved ultimately successful.

After much confusion and many negotiations in the year 2000, the joyous reacquisition of the Lange family's main company building, originally built in 1873, must have given Walter Lange much personal satisfaction, especially since this former Lange manufacturing site is used today to house production rooms and the company's management offices. His great-grandfather received posthumous commemoration when the city of Glashütte renamed the square in front of the train station Ferdinand-Adolph-Lange-Platz and bestowed the number one upon this important and symbolic Lange building. An extensively renovated post carriage signpost from the eighteenth century once again decorates the forecourt of the historical building, just as it did one hundred years ago.

This vital man in his early eighties still fulfills his many representative duties as the "face" of the A. Lange & Söhne brand with great enthusiasm, even if travel to foreign countries has become somewhat more arduous since the death of his beloved wife at the beginning of 2002. In her stead, son Benjamin has become an attentive companion who, in following in the footsteps of his ancestors, has meanwhile become head of the service marketing department. ■

Walter Lange

Time and Again

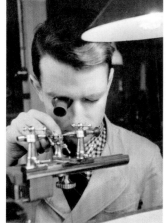

Walter Lange, great-grandson of Ferdinand Adolph Lange, spiritual father, initiator, and founder of today's Lange watches, talks about the "dark years" between the glamorous era of pocket watches and the contemporary horological success story without equal after Germany's political reunification.

Mr. Lange, the golden era of the pocket watch was irretrievably over with the outbreak of World War I. What type of relationship do you have to the very fine pocket watches of A. Lange & Söhne, especially in light of the fact that you weren't even born during their golden age?

■ **Walter Lange:** I got to know these watches very well through my work as a watchmaker in the repair department. You are, however, correct—I am a child of the twentieth century. At the beginning of the 1920s the OLIW watch was developed, a watch that was already outfitted with exchangeable individual parts.

By comparison, almost all of the parts were fitted by hand into the old pocket watches. I remember the pinion turners, how they sat there at their bench lathes with treadle wheels and turned every single pinion by hand, needing a great deal of experience and feel to make them. Those were different kinds of steel back then, and they could not be worked in an industrial manner, but only with a good old LeCoultre graver, and everything was lubricated with Keuper's bone oil. I have a deep respect for the old masters, but their methods back then had no future.

The grand era of fine pocket watches took place between the founding of the Reich in 1871 and the outbreak of World War I. The German upper classes were created, the most important clientele for Lange. Our watches were expensive; it was all about quality not production numbers. The extra-flat hunter's pocket watch that I got for my twenty-first birthday bears a production number that was still under 100,000—and it was manufactured seventy years after the founding of the company!

When did you enter your family's business?

■ **Walter Lange:** After returning from the Russian front in April 1945, I worked in the company for a few months, mostly helping to rebuild. Literally, during the last few hours of the war, on May 8, 1945, our main production building burned down due to a bombing. Only the turbine for the generation of electricity remained untouched. In the fall of 1945 classes began again at the German School of Watchmaking, and I ended my training there under Alfred Helwig in 1947. It had been interrupted by military service. Afterwards, I worked in various departments of the company.

Until the end of the war, we had produced Caliber 48 for use in deck and pilot's watches, and after that the movement was also used in marine chronometers. After the war we literally had nothing. My Uncle Otto, the dominant senior member of our family, wanted to start up again with pocket watches, with the OLIW movement.

I was very adamant about wanting to develop a wristwatch and a wristwatch movement. Caliber 28 was created, fundamentally a smaller version of Caliber 48. Unfortunately it didn't go into production until after the expropriation in 1949. From 1949 until 1951 our watches were made under the name Lange VEB Glashütte, and after 1951 under GUB Glashütte/Sa. The watches received the GDR quality symbol Q, and until 1957 a total of 22,900 pieces were manufactured.

With regard to the expropriation—did this bring additional reprisals for you along with it?

■ **Walter Lange:** After the expropriation I was allowed to continue working at the company, but I was supposed to become a member of the "Free German Union" (FDGB Freier Deutscher Gewerkschaftsbund). I had them show me the statutes in order to be informed about the conditions and responsibilities of membership. At the very top it read that the member had to consent to all of the decisions of the board of directors. I said to them, "You know what happened here. Two local union members and the local SED representative came and declared the company impounded. I'm sorry, but I cannot join." The three owners refused to sign an impounding decree, and from that minute on, they were not allowed to set foot in the company again. I was summoned to the employment office and subsequently sent to work in the uranium mines. On November 15, 1948, I was supposed to show up for "work." I made the decision not to go and fled to the West instead.

The date of the expropriation, April 20, 1948, became exceptionally important for me for a tragic reason after the reunification. In the reunification contract, it was stated that all of the expropriations taking place between the end of the war (May 8, 1945) and the founding of the GDR (October 7, 1949) could not be given back to their rightful owners. The properties expropriated during this lawless period were awarded to the West German Treasury, and the Treuhand, a subdivision of the Ministry of Finance, was entrusted with the care of these properties. A group of expropriated owners started a lawsuit at the Federal Constitutional Court that had established the irreversibility of the expropriations in its decision of April 23, 1991. This decision "made in the name of law and order" felt like a second expropriation to me, and one that disturbed me more than the one in 1948 under an illegal law.

All of the properties and production machines expropriated after 1949 were given back. My cousins, the Eichlers, received their Archimedes adding machine factory back in this way, and we were able to buy it from the inheritors in order to put the Lange II factory there.

What happened when you went to the West?

■ **Walter Lange:** I went to Bavaria, to Dr. Kurtz (former Urofa/Ufag director —ed.) as a watchmaker, my only refuge in the West. My parents and relatives were all in the East, and I had to earn my living somehow. Food was still being rationed, and in order to obtain a food card and residence permit, I had to be recognized as a refugee. My path next took me to the north, to Hof-Moschendorf, the infamous camp where judgment was passed on the "zonies"—either recognition or rejection. After a number of years, the first entry permits for the GDR became available for "Soviet Zone refugees," but I didn't dare to visit Glashütte again until 1976. My Uncle Otto had died in 1972, and his housekeeper advised me to take a look at his belongings, especially the archive material, and bring it to safety. These papers were invaluable to me, and you can imagine the nerve it took for me to hide them under a blanket on the backseat of my car and bring them back to the West.

To continue, at first I was employed as a watchmaker with Dr. Kurtz, who was able to create a small company called Glashütte Tradition from one that had relocated. It was there that I met my wife. The year 1950 brought economic downturn with it, and we looked for new opportunities in the watch and jewelry city Pforzheim. My older, in the meantime late, brother Ferdinand had moved to Pforzheim, and we contacted the Swiss company Altus in Biel from whom we had bought movements before the war. At the end of the 60s, though, we were ready to manufacture and market Lange watches in the West. We looked for prospective partners at IWC in Schaffhausen to aid in this plan. A few pocket watches were created bearing

"... only the turbine for the generation of electricity remained untouched."

The end of an era: A. Lange & Söhne's factory was completely destroyed after the bombing on the last day of World War II.

the words, "A. Lange, previously of Glashütte." This engraving also decorated the plates of the high-quality IWC movements, peaceably placed next to the words "Probus Scafusiae." The reason these watches couldn't make it was the market situation.

In your opinion, where would Lange be today if the company hadn't unexpectedly landed behind the Iron Curtain?

▪ **Walter Lange:** That is not a question I can answer at this point. It's impossible to say. The OLIW watch was being manufactured along with wristwatch Caliber 28. In the 50s, the so-called economic miracle was booming in the West, and large production numbers were in demand. Wristwatches were not luxury goods, but were rather classified as an everyday need. The Swiss industry, the Pforzheim manufacturers, and the gigantic Junghans factory located in the southern part of the Black Forest were producing to capacity and could hardly meet the demand. The completely dismantled Urofa/Ufag factory had to restart production from scratch with many difficulties, just as we did. Then the quartz watch hit the market, and the mechanical timepiece was no longer in demand. The general decline of mechanical watch factories is a well-known fact.

"... Almost more than to rebuild my family's company, I above all wanted to bring work to the people here in Glashütte again."

Did you ever dream that such a renaissance of Glashütte and the A. Lange & Söhne brand would be possible?

▪ **Walter Lange:** Certainly not! I had just gotten the company capital of 100,000 marks together in 1990 and had to "borrow" a Glashütte address—Ernst-Thälmann-Straße 7—from an elementary school chum to register the new company. That was a gigantic act of confidence on her part, for the residents of the GDR had absolutely no experience with such formal legal tricks. When we were looking for employees for our company by putting up a sign at GUB, interested parties literally only dared to come to us after dark or meet us on neutral ground, just like in a spy film. To voluntarily want to change jobs had something forbidden about it. And we were also a bit unsure at the time and called the company Lange Uhren GmbH instead of A. Lange und Söhne due to the not-quite-clear rights to the name at the beginning.

And when I look around now and see how wonderfully everything here in the city has developed, I get a warm feeling all over. There have never been as many people working in the watch industry in Glashütte as today. The mayor says it's between 700 and 750 people. The oft-quoted number of 2,000 employees from earlier on was referring to the entire combine, and the watch company (GUB) also manufactured other things such as calculator movements and precision engineering mechanisms. A maximum of only about 250 people were occupied with just making watches there.

After the fall of the Berlin Wall and the breakdown of the system, contracts were not renewed, and the number of employees was massively reduced. That reminded me of the hard times experienced during the world economic crisis of 1929 when A. Lange & Söhne could only employ its watchmakers on an hourly basis and the unemployed spent their time hanging around on the small bridge across the Mühlgraben behind the train station. As a small boy, I saw the men leaning against the bridge's railing smoking cigarettes, and they seemed so discouraged.

I had to think of that when I came back to Glashütte in 1990. Almost more than to rebuild my family's company, I above all wanted to bring work to the people here in Glashütte again. ▪

Bibliography

The previous chapters concerning the early history of watchmaking in Dresden and Saxony's Erzgebirge mountains and the heyday of the A. Lange & Söhne company were for the most part based on the book *A. Lange & Söhne—Eine Uhrmacher-Dynastie aus Dresden* by Reinhard Meis. This reputable author, who also played a leading role in the conception and realization of A. Lange & Söhne's modern wristwatches, has created a multilayered mosaic containing historical information based on his research in Glashütte and company archives, talks with contemporary witnesses and collectors, technical details, political correlations, and personal memories, making this book the rightful new standard on Glashütte's watchmaking history.

A very moving description of the company's development between the world economic crisis and German reunification is contained in the autobiography of Walter Lange, the company founder's great-grandson. *The Revival of Time* occupies itself with these events in a very personal way, conveying a vivid picture of circumstances before World War II, during the GDR era, and just after the fall of the Berlin Wall.

One of the first books concerning Lange was written by Martin Huber: *Die Uhren von A. Lange & Söhne, Glashütte/Sa.* As a most knowledgeable connoisseur of Lange pocket watches, the author has played a large role in the newfound popularity of the brand, leading Günter Blümlein to comment; "We may have once again made the legend into a watch, but Martin Huber first made the watch into a legend."

There is one more man who has aided the renaissance of Germany's legendary watchmaking city with his books about Glashütte's watches and the German School of Watchmaking, also receiving the key to the city for his untiring commitment: Kurt Herkner, who in the later years of his life even learned to use a computer in order to publish books on his own—as in those days no established publishing house was interested in bringing out works on such an exotic topic. ∎

Reinhard Meis
**A. Lange & Söhne—
Eine Uhrmacher-Dynastie aus Dresden**
(A. Lange & Söhne—
A Watchmaking Dynasty from Dresden in German language)
Third edition, 386 pages
Callwey Publishing
ISBN 3-7667-1272-1

Walter Lange
The Revival of Time
174 pages
Econ-Verlag
ISBN 3-430-15977-6

Martin Huber
Die Uhren von A. Lange & Söhne, Glashütte/Sa.
(The Watches of A. Lange & Söhne, Glashütte/Sa. in German language)
Fifth edition, 216 pages
Callwey Publishing
ISBN 3-7667-0888-0

Kurt Herkner
Glashütte und seine Uhren
(Glashütte and Its Watches in German language)
440 pages, out of print
Herkner Publishing
ISBN 3-924211-05-1

Kurt Herkner
Glashütter Armbanduhren
(Glashütte Wristwatches in German language)
304 pages, out of print
Herkner Publishing
ISBN 3-924211-06-X

Günter Blümlein

The Watch Maker

In the 1999 first edition of this publication we presented an empathetic portrait of Lange's visionary managing director and cofounder Günter Blümlein. Author Manfred Fritz found a wonderful analogy to illustrate the Nuremberg native's thirst for action: His favorite pastime was picking up trains that had fallen off their track and putting them back on their rails. Blümlein picked up three of them during his lifetime, pushing all three locomotives—IWC, Jaeger-LeCoultre, and A. Lange & Söhne—at full speed into the top segment of the watch industry.

Shortly after Blümlein had led his three "children," as he often liked to call the watch brands comprising Les Manufactures Horlogères (LMH), safely underneath the roof of the Richemont concern, his own train derailed, and neither friend nor doctor could get it back on track. Blümlein passed away in October 2001 after a brief, painful illness. He achieved a great many wonderful things in his fifty-eight years, but he certainly would have done a great many more.

Manfred Fritz reviews the life of one of the most charismatic and successful managers of the watch industry.

Günter Blümlein, born in 1943, brought an ideal combination of knowledge and experience with him to A. Lange & Söhne. An engineer in precision mechanics, he worked in the development department of the Diehl concern in Nuremberg, subsequently moving on to test his talent at marketing. From there he transferred to one of the concern's subsidiaries, Junghans, located in the Black Forest, which was at the time the largest watch factory in the world. He took over direction of quality control, becoming deputy director of technology and finally managing director of marketing and distribution. A long career without a blemish.

And then came the end of the 1970s and that which still seems like a bad nightmare to the entire Swiss watchmaking industry: the quartz invasion from Japan that decimated Switzerland's most important commercial sector. The German concern VDO Adolf Schindling AG, wanting to forge a European watch holding company as a counterweight, purchased IWC and Jaeger-LeCoultre as well as a French company. Nothing came of the great European solution, though, and the two formerly reputable Swiss companies were in need of someone to get them back on track. And that someone was Blümlein. Lightning-quick he made IWC into a technical forge that feared the use of neither titanium nor ceramic. Blümlein was one of the first to understand with his almost predatory sense for market trends that the quartz watch as a mass-produced product would soon take its victory to the grave. The men eventually separated from the boys, and high-quality mechanics with and without additional functions were once again the order of the day for the high-end market segment. This he took advantage of in Schaffhausen with unmistakable, emphatically masculine watches, including the most successful calendar watch of all time, the Da Vinci, and the grandiose Grande Complication.

Günter Blümlein

Starting in 1984 the polyglot manager married to a French national also had a standing office at Jaeger-LeCoultre in Le Sentier, commuting every two weeks between Schaffhausen and the Vallée de Joux in trying to buff the sister company to her old shine. This firm had all of the necessary prerequisites: know-how, elite watchmakers, production capacity—only the right products and marketing were missing. Here it was again: Put it back on its tracks and give it a push. Blümlein reduced an adventurously wide product palette, woke up the horological answer to Sleeping Beauty, the Reverso family, added the Master line, and positioned the distinguished brand with its fine mechanical movements as a complement to the technically interesting "men's watches" from Schaffhausen.

Until that point, he had been balancing haute horlogerie with two stars. The third star, shining brightest for him, surprisingly rose in the east in 1990 with German reunification and Walter Lange's founding of Lange Uhren GmbH. With a powerful financial push, the ability to educate watchmakers properly from inside the group, and forty-eight excellent tradesmen in Glashütte, he planned and realized together with Walter Lange the most astonishing success story in recent watch history. For him it was also, as he freely admitted, a "patriotic" challenge. Despite the great name that led fine German watchmaking to world renown from 1845 until 1945, they had to start from scratch in Glashütte. But this challenge was an opportunity for a man with Blümlein's energy and acumen: The direction—the top of the luxury watch market—was decided, but this time new tracks had to be laid.

The results of his toil are visible running the gamut from the Lange 1 model, a watch upon which prizes and awards have literally been heaped, to the Langematik and the tourbillons. And they have been observed with Argus eyes—for he set new heights in quality and innovation and, with clever market strategy, once again positioned the watchmaking predicate "made in Germany" in the world market. Way up high. ∎

German President Roman Herzog honors Günter Blümlein in June 1997 for being one of the "courageous entrepreneurs that our country needs."

The "Delicate Little Flower Called Lange"

The late Günter Blümlein, former managing director and visionary at A. Lange & Söhne, talked to us in late 1998 about the difficult resurrection of the brand name, the concept of the new Lange watches, "typical German" design, international success, and the long-term prospects of watches "made in Germany."

When did "Project Lange" begin for you?

▪ **Günter Blümlein:** In 1989, when the GDR was beginning to disintegrate and the first "republic refugees" had made it through to West Germany via Hungary, the names Glashütte and A. Lange & Söhne were brought up in our group (the watch group of the German Mannesmann/VDO company, of which Günter Blümlein was the president —ed.). How did this happen? VDO's former supervisor of the board of directors, Albert Keck, was a watchmaker himself. His instructor had come from the German School of Watchmaking, and IWC had also already had contact with Walter Lange, the great-grandson of Ferdinand Adolph Lange, in the 1960s, planning a cooperative venture to manufacture Lange watches in the West. This was doomed at the outset because a brand like A. Lange & Söhne is dependent on the collective spirit of the region's people. It is impossible to just plant a foreign movement in a Lange, even a Swiss one.

At the beginning it was more a patriotic feeling that caused us and Walter Lange to present our ideas for the region surrounding Glashütte to the former man-

> *"... and it was very clear to me that the delicate little flower called Lange could only thrive if it was set free."*

ager of the large combine of microelectronics and precision engineering in the fall of 1989. At the time, this combine comprised Robotron, which was made up of diverse computer chip and calculator manufacturers as well as the watchmaking companies in Ruhla and Glashütte (GUB), employing a total of about 40,000 people. Our idea was to go into a joint venture with the existing eastern watch group, bringing in our know-how. We were one of the very first to show serious and motivated interest in committing ourselves industrially to East Germany.

In February of 1990 we were able to visit Glashütte, or more precisely GUB, and we sobered up quickly. What we had already put into our concept was confirmed with just one glance. There was no question: A. Lange & Söhne would never be able to rise from this much-too-large VEB (Volkseigener Betrieb or Company of the People) combine in a joint venture. Lange had to be an independent company, quality and market oriented, free from material and spiritual baggage—the possession of which would make a breakthrough to the top of the luxury watch industry unthinkable.

The combine managers weren't interested in our musings: if we wanted to take over, we had to take over the entire kit and caboodle in Glashütte, including its nearly 2,000 employees. Don't forget that this was a good six months before the reunification even happened and there was no way past the combine management. They were interested in cooperating with us and also wanted to see Mr. Lange fulfill his "historical obligation" as they put it. I was well acquainted with companies of this size and their gigantic production numbers from my own experiences (Günter Blümlein had been managing director of Junghans for many years –ed.), and it was very clear to me that the delicate little flower called Lange could only thrive if it was set free.

The joint venture idea was overtaken by the surprisingly quick completion of German reunification, and Walter Lange became active, founding his Lange Uhren GmbH on December 7, 1990. Our watch group took a majority share in the company, contributing a substantial increase in capital stock. The company's initial address was Ernst-Thälmann-Straße 7, the residence of one of Walter Lange's school chums.

Restart after Reunification

"For the watchmakers of Glashütte, Patek Philippe, 1,000 kilometers south in far-away Geneva, was not the measuring stick."

We started this company from nothing! But maybe it was the catastrophic surroundings, the ground zero, which gave us the courage and wings to create what we presented to the watchmaking world four years later, something they received with ever-growing astonishment.

What happened after the company was founded? You had nothing in Glashütte, no factory, no employees . . .

■ **Günter Blümlein:** We looked for our first employees by hanging signs on the bulletin board at GUB, a situation that was okay with everyone involved, for outside of this large company there was no one else to be found in Glashütte. As all the old GDR markets broke down, GUB's business was floundering, and personnel had to be drastically reduced, as was also the case in the four years between 1991 and 1994. This huge concern was finally pared down to seventy people, who have worked in a new version of the company since 1994, managed very successfully by Heinz Pfeifer.

But back to the year 1990 when we wanted to build up our workforce. You shouldn't think that gobs of people came to us! Most of GUB's employees were still hoping for a "complete solution," someone who would take over the entire company as it was. We seemed fairly exotic to them with our dream of a comeback for A. Lange & Söhne—which is understandable if you will recall that there was nothing like marketing in a Western way in the GDR. GUB had been the professional home of two generations of Glashütte residents. The people there viewed free entrepreneurship à la Lange as the "enemy of the working class."

We had about 120 applications for the first round of employees needed, from which we chose a good dozen of the most excellent trade professionals. There were many candidates among them whose grandfathers had worked at Lange. One of them even brought a little box with him to the interview in which carefully wrapped parts of a tourbillon that his father or grandfather had started back then were stored. There was another designer who spontaneously applied with Walter Lange out on the street and introduced herself with the words, "I am the granddaughter of your former chronometer maker, Mr. such-and-such." That was naturally enough reference for Walter Lange to recommend the lady to us heartily.

The combine management back then, and later also the Treuhand, welcomed our efforts in creating highly qualified jobs. Since all of the workshop space at our disposal in Glashütte belonged to the combine at that point, the main Lange building was vacated for us and prepared to fit our needs. Hartmut Knothe, back then still director of the central watchmaking school called Makarenko that belonged to GUB, was our first employee and had his office at Altenberger Straße 1.

Just a moment—Lange Uhren GmbH was located at Altenberger Str. 15, the former precision clock factory building belonging to Strasser & Rohde, wasn't it?

■ **Günter Blümlein:** That is correct. But in 1991 Treuhand jurisdiction had just changed from Thuringia to Berlin, and all of a sudden the offer of letting us have the Lange main production building including its archive and museum was retracted. We were out on the street again. Lange then found a worthy home in the production building formerly belonging to the precision clock factory Strasser & Rohde.

The space problems certainly weren't the only ones you had to deal with. Did you not have trouble with GUB and the Treuhand due to brand rights?

■ **Günter Blümlein:** That's a long story. Looking back, until the spring of 1993 we had a fairly undignified squabble with the Treuhand and GUB concerning the rights to the name and brand, actually leading to a few legal clashes. You have to keep in mind that all trademarks from former Glashütte watch manufacturers were no longer permissible according to the new brand law West Germany passed in 1993. But it was in no one's interest to hand Glashütte's future over to judges and lawyers. Glashütte needed to belong to the watchmakers, and that's the whole reason we went back there.

To make a long story short: We settled with GUB concerning the rights to the brand in a written contract. For example, we agreed that Lange could refer only to the former watch manufactory A. Lange & Söhne and its tradition of demanding horology. We also agreed that GUB, which had become Glashütter Uhrenbetrieb GmbH in 1991, could refer to the founding date of 1845 in its

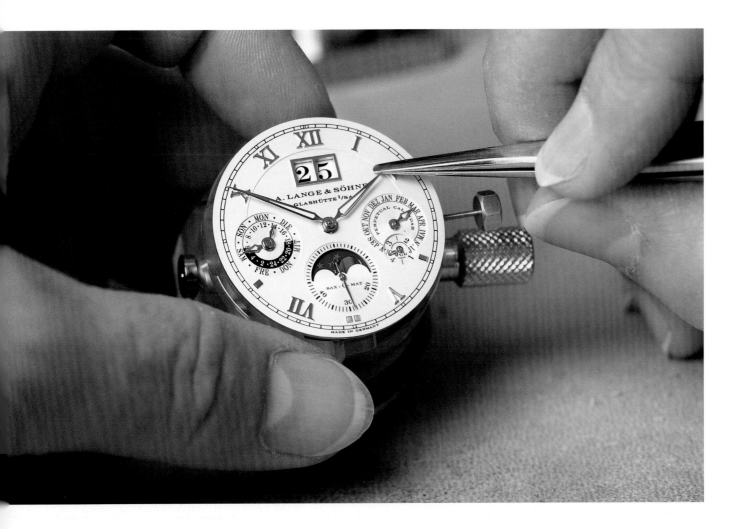

advertising. Our Walter Lange had a hard time swallowing that, of course, because 1845 was the year of Lange's company founding, the first industrial production of fine watches in Germany. And after World War II, this family business along with six other medium-sized concerns was unlawfully expropriated by the SED regime in order to turn them into the newly founded VEB Glashütte in 1951. The rightful succession of this "people's company" should have been granted to the oldest of these expropriated companies—the descendants of the expropriated must feel that this was an awful injustice. But in negotiations there is always a give and take, and unfortunately German history cannot be rewritten.

In not giving back expropriated properties, the injustice of the SED regime was judicially cemented, but the government was trying to keep it simple in many areas. Because of this, it became impossible to transfer the old A. Lange & Söhne company back to the remaining members of the Lange family. We are mainly talking about Lange's main production building, the company's archives, and the museum. These things later became the property of the city of Glashütte.

Today we can all concentrate on things that are really important in Glashütte—the development and manufacture of fine Glashütte timepieces. The general success of the region should soon help to heal all wounds.

> "Glashütte needed to belong to the watchmakers, and that's the whole reason we went back there."

Why didn't you want to take over the entire GUB operation? Then you wouldn't have even had all of these problems.

■ **Günter Blümlein:** Watches like we wanted to make, in the vintage Lange manner, had had no tradition at GUB for over forty years. Where would we get credibility? From our point of view, the rebirth of the A. Lange & Söhne brand was only possible if we remained separate from the baggage of the inflexible VEB combine.

In addition, I have always been for healthy competition here in Glashütte; it corresponds to one of my business

"The brand A. Lange & Söhne is something like a collective work of art ..."

principles. All brands belonging to the same company? Glashütte's residents had had enough of that, and it was not our goal to monopolize the place. Competition, or as Walter Lange likes to put it using the words of his great-grandfather, "the Lange company as an alum crystal, around which new things are built," is the mainstay of this watchmaking region. For the watchmakers of Glashütte, Patek Philippe, 1,000 kilometers south in faraway Geneva, was not the measuring stick. Other sources of motivation and adrenaline were needed here—competition within the city. With this vision in front of us, Lange hit the market in 1994 with the most luxurious of watches and practically prepared the earth as would a plough for the success that all of the companies now located in Glashütte enjoy. This is contributed to by all of the companies located in Glashütte, be they Lange, Glashütter Uhrenbetrieb, Nomos, Mühle, or Union. Naturally competition plays an important role. Thus, we do not regret not having taken over GUB back then.

Did the international success surprise you?

■ **Günter Blümlein:** I'll admit that the way it has developed has. Of course, we dreamed of this, but generally we expected to be successful with a "German luxury watch" first in Germany, and less successful at that. With a certain irony I can admit today how we wanted to prepare ourselves "optimally" for the home market: The Cabaret model was supposed to be called "black-red-gold" (the colors of the German flag —ed.), for its black dial, decorative red seams on the strap, and gold case reminded one of this. And where today the word *Doppelfederhaus* ("twin spring barrels") is placed on the dial of the Lange 1 model, we were originally going to put the word *Deutschland*. The first prototypes looked exactly like this. But the strong nationalistic feelings following

German reunification were quickly replaced by everyday ones. Normalcy was once again the order of the day, and these curious ideas, as they would be regarded now, disappeared along with those feelings.

Strategically speaking, we didn't want to make A. Lange & Söhne a domestic brand. For this reason, we created our distribution policy right from the beginning to maintain a balanced relationship between domestic sales and export. Already six months after A. Lange & Söhne's comeback, we opened select doors of fine jewelers in Italy, Southeast Asia, Japan, the U.S.A., France, and Great Britain. Watch connoisseurs the world over literally pulled us into their markets, and the international trade press certainly had a lot to do with that. Soon the ratio of domestic to foreign was already 60 to 40 percent. International success was certainly a strategic part of our plan, but this amount was also surprising to us.

Let's get to the main topic: A. Lange & Söhne's watches. These timepieces don't deserve special attention only because of their success in the marketplace—they would make a good impression anywhere, anytime, due to their style and technical excellence. How did it come about that the modern-day Lange watches look the way they do and are the way they are?

■ **Günter Blümlein:** Lange watches represent the quintessence of our feelings and experiences; they must express the content of the brand A. Lange & Söhne physically. And the brand A. Lange & Söhne is, if you'll excuse the expression, something like a collective work of art, comprising a long history, employees with a passion for the fine watch, the style of the house, the obligation to tradition, and last but not least the unique technology and craftsmanship prescribed by Lange itself.

Restart after Reunification

Hand engraving on a solid case back.

All of this, putting these ideals into a shape, into technology, wasn't and isn't easy. We have left nothing to chance. One of the reasons for the homogenous design of our Lange collection was certainly the fact that these initial watches came from the same place and thus seem like they are cast from the same mold. Although you don't have to take the words "the same place" literally. With the rising number of participants needed in light of the meteoric growth of our company, diverse ideas of technology and design will also arise within our walls. The battle for the authentically styled, unmistakable, simultaneously understated and simple Lange watch will always have to be fought anew. And now with the growth of the company, it will make this task more difficult than it was when we started from scratch.

Our Lange watches were to be "typically German," different in their design and charm. And it's this difference that forms the elementary core of the design. I personally enjoy the task we have set for ourselves at Lange – being different than the Swiss – because I have already had the pleasure of taking part in the creation of many successful Swiss watches. Thus it is easier for me to define what needs to be different and what can be different.

Everyone makes an effort to be "different" . . .

■ **Günter Blümlein:** Well, in every industry there are those who wish to reach a position for them-

"... in the end, it is a Lange's inner values that cause it to be purchased."

selves with innovation and those who search for their success with imitation. It is my opinion that a large part of the watch industry is unfortunately composed of imitators. I would never, for example, use such a showy element in a watch as a diamond-cut bezel because Rolex already does this. Or look at how quickly the entire industry rushed to imitate the Lange outsize date. Stealing ideas seems to be the fastest way to the finish line for most. Wanting to be different is also a question of personal attitude, of character.

With Lange we were very careful not to look anywhere else. We sat down and thought about what could be "typically German" about a watch. This is how we arrived at the weight of the watch, for example. Imagine the sound that closing the door of a fine German luxury automobile makes, for example. It makes a "plop" sound, and customers like that; it is perceived as a symbol of value. When I hold a Lange watch in my hand, its healthy weight also somehow causes this "plop" sound to go off in my head. But that's only one small example.

I am always a little embarrassed when I use the words "collective work of art" for a Lange watch. But everything has to be harmonious on a Lange. It is the nature of the beast that the first serious attacker of the Swiss bastion of the highest-priced luxury watches would especially stand in the spotlight and be critically observed. For this reason, a Lange is designed all the way. Even if you turn the watch around, this feeling of Lange value will jump right out at you. We use just as much care in the design of the movement and its individual parts as in the design of the case, dial, hands, and strap buckle. In the end, it is a Lange's inner values that cause it to be purchased. I will even go a step further: If you understand design as an especially conspicuous element or even striking styling, then our watches fundamentally have no design.

I beg to differ: The design of a Lange watch may not be visible, but it is something you can feel. Only a small percentage of buyers make the decision to buy technical quality.

■ **Günter Blümlein:** Correct, you don't see the design, you have to feel it, like you said. And people may not understand the technical quality in every case, but they should feel that as well. Fifty percent of our product design takes place in the movement. Our engineers have to downright haggle with the designers over the look of the individual parts, for very early on, to an extent before the actual development and design work even take place, a separate design booklet is created for the movement designer alongside the booklet for the product. I would say that this makes us different from most others in the industry—in our case, technology follows design. That's why our movements are always custom-made, fitting exactly to the shape and size of the watch's case. And this is only possible if you make your own movements; there are no standard solutions. The movement design of the Lange 1 model or that of the Datograph is fundamentally crazy constructions, completely unconventional in the arrangement of certain important parts like the winding shaft, the spring barrels, the position of the hands on the dial, and the date display. This open-minded thinking in the design of the movements is certainly one of the most outstanding characteristics of Lange's contemporary art of watchmaking, and it most certainly sets us apart from our competitors.

You are well known as a perfectionist in the industry when it comes to design. How do you see this?

■ **Günter Blümlein:** In questions of design, if you align yourself with terms such as timelessness, understatement, and refined simplicity and refuse applied style elements and decoration entry, then only the details remain, and they need a lot of time. We wrestle and haggle with every tenth in order to get a just barely perceptible nuance in the color. The shapes and edges are harshly judged and criticized with fingertips. Even the nose and ears take part, when I think of the smell of leather straps, the sound of the movement being wound, and the activating of the chronograph mechanism.

The Langematik, for example, took us a lot longer than the Lange 1 model, although it is not as conspicuous a model as the Lange 1. As an automatic watch, the Langematik was consciously designed to become something different than the Lange 1. It is always difficult to find a worthy successor or a worthy addition to something that is already very special, like the Lange 1 indisputably is, especially since this model became the symbol for our brand right at the beginning because of its seemingly exotic asymmetrical dial division. This all didn't make designing the Langematik any easier; this watch was also to be simple like all of our creations. I guess it was necessary to become a perfectionist.

"... our engineers have to downright haggle with the designers over the look of the individual parts ..."

What do you think Adolph Lange would say if he could see the contemporary Lange watches?

■ **Günter Blümlein:** Well, I could imagine that he might ask in an astonished way why we would want to squeeze his high-quality Saxon watchmaking into such a small case so that we can wear it on our wrists. (Adolph Lange died in 1875, a quarter of a century before the wristwatch became popular –ed.)

And, finally, perhaps you could summarize in a few words how products are developed and manufactured at Lange.

■ **Günter Blümlein:** You know one of our advertising slogans, the one that we generally use outside of Germany, is: "The Swiss make the world's best watches. So do the Germans." We want to—and this is how Walter Lange and our company had it put down in our partnership contract—construct the best watches in the world once again, in the quality and tradition of the legendary pocket watches of the old Lange manufacture, always asking ourselves, how would Ferdinand Adolph Lange have thought and acted if he were still alive today and had to make a wristwatch.

Without having known Ferdinand Adolph Lange personally, I believe that you are acting as he would have done. Thank you very much for the talk. ■

(Peter Braun interviewed Günter Blümlein in December 1998.)

Project "A. Lange & Söhne"

A Legend Becomes Time

The first public appearance of the new Lange watches on October 24, 1994, sent out a worldwide signal: Something special had been created from literal ruins—a phoenix rising from the ashes. This something was so different from everything else that it was apparent at first glance; this something also self-confidently passed the scrutiny of a second and third glance. The tangible and verifiable excellence of the watches proudly bearing the name A. Lange & Söhne on the dial revived a 150-year-old legacy and enriched the horological world with the addition of a remarkable legend.

Snapshot from fall 1994: Hartmut Knothe, Walter Lange, Günter Blümlein

Project "A. Lange & Söhne"

It wasn't even four years after the new registration of the brand name A. Lange & Söhne that the rumors surrounding the "German luxury watch" verified themselves in the shape of a cleanly structured collection comprising four timepieces: Lange 1, Saxonia, the women's model Arkade, and the tourbillon Pour le Mérite.

The reconstruction of the "new old" brand, the development of the product, the furnishing of the production rooms, and the training of employees constituted a race against time. No one wanted to predict how long the euphoria surrounding Germany's reunification would last, but it was clear that this mood had to be utilized in order to quickly collect some displayable success.

Focusing on the absolute top segment of the watch pyramid, the company had an ambitious goal. Initial investments of 20 million German marks (approx. $10 million today) were on the line, and a large amount of state financial support caused Lange to have to validate itself in the eyes of the Treuhand as well. The economy of Germany's new states needed a *succès d'estime* just as urgently as those in charge of the project did to justify it to the Mannesmann/VDO watch group who, with International Watch Company (IWC) in Schaffhausen and the manufacture Jaeger-LeCoultre in the Vallée de Joux's Le Sentier, already had two flourishing watch companies in its portfolio.

Walter Lange's voluntary inheritance proved to be a heavy load to bear, for *haute horlogerie* circles expected watches bearing the predicate "Lange, Glashütte" to be nothing short of remarkable—something wholeheartedly promised a bare few weeks after the company was founded to the fans of German watchmaking, a crowd that had been starving for Lange for the past forty years.

The Horological World Holds Its Breath

The result presented on October 24, 1994, surpassed all expectations: Lange's watch philosophy had been miniaturized and transported into the modern era. It doesn't matter that it happened seventy years later than it should have—the successors to Lange's third generation proved that the elementary virtues of high-quality pocket watches can be carried over to wristwatches. Thus four "virtuous" watches were created that would undoubtedly have made founding father Ferdinand Adolph Lange proud. The outside elegantly simple, though not insipid, and the inside the most beautiful mechanism that Glashütte traditions have to offer, outfitted with a polished three-quarter plate made of radiant German silver, screw-mounted jewel chatons, and numerous technical

When the name Lange is mentioned, something remarkable is of course expected ...

niceties that once created the legendary reputation of Lange's pocket watches.

Lange's workforce, initially comprising forty-eight employees from the surrounding region under the leadership of former vocational school director Hartmut Knothe, was put through an additional training session in modern production technology at IWC in Schaffhausen in order to be able to operate the highly precise CNC machines, some of which were built especially for Lange Uhren GmbH. Concerning the craftsmanship skills of the young watchmakers, designers, engineers, and toolmakers, the Swiss "instructors" were surprised at the high amount of technical education the young men and women had received in former East Germany. It was apparent that the Lange team's motivation accelerated the transfer of knowledge.

However, it was still necessary to learn how to handle the responsibility for the company's own history, for no one in Glashütte except Walter Lange, the company founder's great-grandson, remembers the "good old days." A weekly visit to the Glashütte watch museum, part of the teaching curriculum in a way, sharpened the eye for fine horological details lost in forty years of VEB large-series production.

During the four-year preparation period, the term "German luxury watch" took on a tangible and visible shape, fulfilling both new and old criteria for the art of Glashütte watchmaking, going step by step to rehabilitate the characteristics unchanged in more than one hundred years of legendary Lange quality. ■

The Watches of A. Lange & Söhne

Lange 1

The Success Story

After its launch in 1994, the Lange 1 model quickly became a favorite of watch fans, advancing to a symbol of the manufacture A. Lange & Söhne and, as such, the craftsmanship of all Saxon watchmakers.

It would not be an exaggeration to say that the Lange 1 model, inundated with prizes and awards since its introduction, wrote a chapter in horological history. This watch is remarkable not only in its outer appearance and trailblazing technology; perhaps the most remarkable thing about this timepiece was that in the year 1994 no one expected the watches of the "new" A. Lange & Söhne to continue so seamlessly where the watches of the "old" A. Lange & Söhne left off, as if the forced break of almost fifty years during the thankfully transitory existence of the first German classless state had never happened.

The pressure of the expectations resting on the shoulders of Glashütte's watch engineers was enormous, for the ancestors of Walter Lange had placed the bar high. And the record had to be beaten. Anything less would have been commented on by trade and watch enthusiasts alike with malicious delight.

It has become clear, however, that the Lange 1 model is the flagship of the new A. Lange & Söhne brand, setting especially high standards, almost as if to say to the horological world, "See what we can do."

The models Arkade, Saxonia, and especially the tourbillon Pour le Mérite, also introduced in 1994, made it abundantly clear that this quality was to be the standard for all watches made by A. Lange & Söhne. Suddenly a new light appeared in the horological sky—or was this a comet, appearing on its preordained orbital course

Lange 1

The Lange 1 is available in many variations of yellow, white, and red gold as well as platinum: diameter 38.5 mm; height 10 mm; sapphire crystal; silver dial; gold hands; gold crown; sapphire crystal case back or solid gold or platinum.

Lange 1

The Lange 1 (38.5 mm diameter) and the Grand Lange 1 (41.9 mm diameter) also differ in their dial designs.

again after fifty years, making some of Switzerland's watch stars suddenly seem like less important starlets? In the past decade it has become obvious that the heavenly body is no flashy comet but is actually a true fixed star. It's good to know that astronomers are also sometimes wrong.

What is the fascination that the Lange 1 holds, even for those who are not absolute Lange fans?

One of the reasons for the success of A. Lange & Söhne watches and the brand's excellent Lange 1 model seems to be that they are manufactured in Glashütte with a great deal of effort using long forgotten ingredients—ingredients that others no longer include today because they are just no longer necessary, such as gold chatons and swan-neck fine adjustments.

In general, the high-quality mechanical watch and in particular, the Lange 1 give the owner a feeling of pleasure, caused by a wonderful little machine existing in our sober, success-oriented world without, as is usually the case, causing any worries about the goal-oriented usefulness of the object.

Günter Blümlein, back then president of the LMH Group (Les Manufactures Horlogères), to which, in addition to A. Lange & Söhne, both IWC and Jaeger-LeCoultre belonged, once commented on this subject in an interview broadcast by the Deutsche Welle radio program:

"Before the war Lange watches were always handmade—very highly developed, cultivated, individual pieces. We are following in this tradition, making very small series, individual watches. Every movement is assembled by hand and contains parts engraved by hand without the aid of lasers, making them unique. We have some very anachronistic components in our movements, such as the three-quarter plate. This component used to have a reason for being. Today, movements are so highly developed that elements such as these are no longer necessary. But everyone who sees them loves them!"

Lange 1

The Strength of Two Hearts

The extremely long power reserve of the manually wound Lange 1 is based on a design incorporating two serially operating spring barrels. The mainsprings inside these barrels are wound simultaneously when the crown is turned. When the crown is turned on other watches with twin barrels, the ratchet wheels of both barrels are turned simultaneously by the winding gears. Not so with Caliber L901.0: here only one ratchet wheel is connected to the winding mechanism. The ratchet wheel of the second barrel is not actually a part of a click, for even though it is located on the square hole of the spring's core like a ratchet wheel, it does not have a pawl. Lange therefore christened this wheel the "spring barrel wheel."

How Does It Actually Work?

Lange 1's crown is used to wind the watch. The winding gears turn the ratchet wheel located on the first barrel. The click is pushed between the teeth of the ratchet wheel by the click spring. The interplay of these three parts assures that the mainspring can build tension.

The first barrel does not have a direct connection to the movement, but is connected to the "barrel wheel" (ratchet wheel) of the second barrel via a small intermediary wheel. The "barrel wheel" of the second barrel and its spring core begin to rotate using the energy that is lent the first mainspring when it is wound. The second mainspring is thus wound by the first, whereby the first mainspring continues to let out tension until the second is completely wound—a simultaneous operation. Finally, both springs are fully wound, and spring barrel number two can power the movement. Clever switching of the barrels and connecting pinion allows the mainspring of the second barrel to relieve a bit of tension before it receives more power from the first barrel.

Because of this construction design, both mainsprings need varying amounts of barrel revolutions to release their tension: about 7.75 for one and 6.2 for the other. These two barrel revolution numbers have to be added together since the spring barrels work serially.

The transmission ratio of the minute wheel that revolves once an hour is 1:7.6, from which the autonomy can be calculated: 7.75 plus 6.2 equals 13.95 barrel revolutions. The transmission ratio says that we have to multiply this by 7.6, equalling a theoretical power reserve of about 106 hours, or a good deal over four days.

That's It in a Nutshell

The Lange1 is hardly turned over to take a peek through the sapphire crystal case back when a crazy feeling of enthusiasm overwhelms the observer. The enormous plate made of German silver and decorated with Glashütte ribbing is reminiscent of a prehistoric animal's armor, protecting the sensitive parts ticking within. The busy balance swinging away at 21,600 vibrations per hour, the escape wheel pulsating nimbly underneath its bridge, and the nine gold chatons in which the gear wheels' jewels are placed all leave the impression that exciting micromechanical processes are taking place under the plate.

The deep blue screws with which the complicated little machine and its individual parts are held together are incredibly pleasing to the eye. The polished endpiece on the upper escapement wheel pivot and the delicate, yet strong, swan-neck fine adjustment on the hand-engraved balance cock make the movement a feast for the eye.

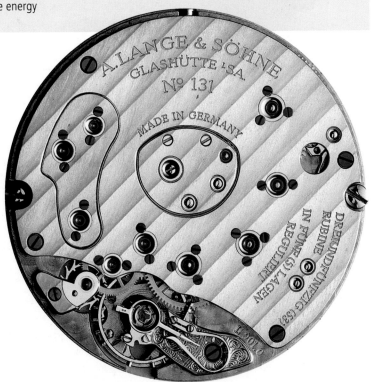

Lange Caliber L901.0 has a diameter of 30.4 mm and is 5.9 mm high. Three-quarter plate made of German silver with Glashütte ribbing; nine screw-mounted gold chatons; hand-engraved balance cock; swan-neck fine adjustment; stop-seconds. The movement is regulated in five positions, and its glucydur screw balance makes 21,600 vibrations per hour.

Lange 1

This dial changed the world of watches while remaining extremely versatile. At top a jeweled version of the Lange 1 featuring a mother-of-pearl dial and a stingray strap. Below, the Lange 1 Luminous with its sporty, contrasting dial.

A Distinctive Dial

Contrary to a "normal" watch, the Lange 1 possesses an off-center display for hours and minutes—the result of an eccentric movement design. Lange 1 is outfitted with Caliber L.901.0, whose minute wheel rotates not as usual in the middle of the movement, but rather in an off-center position.

Some distance away from that, the hand on the dial's edge points out the passing seconds. In order to make this possible, an additional wheel attachment was created for the fourth wheel within the actual movement. Having the fourth wheel carry the little hand itself, as it normally would, wasn't possible because of the unusual dial divisions.

The large hand for the power reserve display, the length of which is close to that of the minute hand, located right above the subsidiary dial for the display of seconds, finds its way from AUF to AB in about three days' time. The main movement design trait of the Lange 1—its great power reserve—was achieved by constructing two spring barrels whose springs release their tension serially, thereby tripling the usual autonomy of Caliber L.901.0 as compared to other manually wound watch movements.

The outsize date window, the hallmark of A. Lange & Söhne's modern timepieces, is very conspicuously positioned above the power reserve hand. This is a design element that has meanwhile oft been imitated.

Today this award-winning watch is available in many variations. Alongside the original case size of 38.5 mm diameter, a sweet little version bearing the nickname "Little One" is also available, especially conceived for more delicate wrists, for which reason versions that include gem-set bezels are very popular. The movement fills the entire space of the 36 mm case without leaving room for a movement holder ring or even the characteristic square correction button for quick-setting the outsize date. Instead, "Little One" bears a corrector recessed into the case that is activated with the pointy stylus that comes with the timepiece.

Space problems are something that is unknown in the Grand Lange 1: with a proud 42 mm case diameter, this timepiece perfectly fits the current taste for larger watches. The scales of the minute and second rings—"subsidiary seconds" is really the wrong term here—were increased in size as well, with the window belonging to the double-digit outsize date display jutting its lower left corner into the hour ring. This case size is the only variation of the Lange 1 to include interesting dial versions comprising contrasting scales.

Otherwise, the classic combinations of yellow gold with a champagne-colored dial, red gold with a silvered or anthracite grey dial, white gold with mother-of-pearl or an anthracite grey dial, and platinum with a silvered or black dial have become the favorites of the collection. ∎

LANGE 1

Manufactured since	1994
Movement	mechanical manually wound movement, Caliber L 901.0, diameter 30.4 mm, height 5.9 mm, 53 jewels, 9 screw-mounted gold chatons, glucydur screw balance, Nivarox balance spring with special terminal curve, swan-neck fine adjustment and patented beat regulation, twin serially operating spring barrels, Glashütte three-quarter plate, bridges decorated with Glashütte ribbing, stop-seconds, 21,600 vph, power reserve 72 h
Functions	hours, minutes, subsidiary seconds, power reserve display, large date (patented)
Case	diameter 38.5 mm, height 10 mm, sapphire crystal, sapphire crystal case back secured by six screws, corrector button for date display on case edge

Direct Flight to the Moon

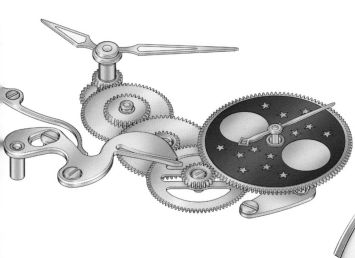

In 2002 the Lange 1 model family once again received an addition, though this time it was in the shape of an especially precise moon phase display. From the mechanically defined deviation of not more than 1.9 seconds per day, a total of a mere 57 seconds' deviation results per moon phase. Only after a period of 122 years and a good seven months can one expect to see a lapse of an entire day.

The design of a moon phase display demands a great deal of mathematical talent, for the lunar month is fairly uneven: 29 days, 12 hours, 44 minutes, and 2.9 seconds. To simplify the process, most movement manufacturers base their calculations on a cycle of 29 and a half days, and if not only one, but two, lunar cycles are deposited upon the moon display disk, one actually comes up with an even 59 days, cleanly divisible by 24 hours, and thus a simple result. Along with this simplification, imprecision is, of course, preprogrammed, for two lunar months in reality do not equal 59 days even, but rather 59 days, 1 hour, 28 minutes, and 5.8 seconds.

Contrary to usual displays, the moon phase gears of this Lange 1 model do not get their energy from a finger switch, but rather—via several intermediary wheels—

The Lange 1 Moon Phase in a platinum case. It is also available in yellow and red gold. Diameter 38.5 mm, height 10.4 mm; silver dial; gold hands; gold crown; sapphire crystal case back.

Lange 1 Moon Phase

The dial side of Caliber L901.5: The correction lever for the moon phase display is visible to the lower left. Above that, the hour and minute stems jut out from a drilling in the dial's plate, while the power reserve display's stem is located near the date bridge.

Explosion illustration of the outsize date and moon phase mechanisms located between the base plate and the dial. They contain a total of ninety-three individual components.

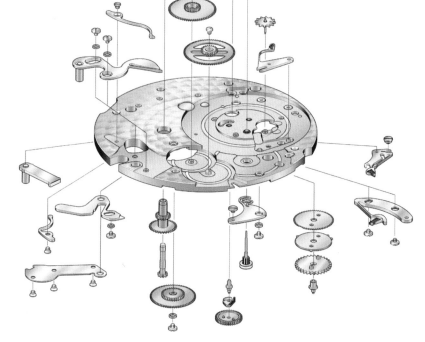

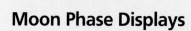

Moon Phase Displays

The moon, feminine in Romance languages (but not in German, an Indo-Germanic language, where it is masculine) and revered in many antique cultures as a symbol of fertility, has always fascinated mankind even without tending overly toward the mystical.

The moon and its continuous changes—from wispy crescent to proverbial man in the moon, then disappearing again to make a new moon—have attracted clock- and watchmakers from the beginnings of timekeeping. Displaying various lunar appearances on a clock became a favorite pastime of many, especially due to the regularity of this natural occurrence.

This partiality to moon phase displays was surely founded in practicality, for especially in the moderate climatic regions of our earth, often clouded over, a glance to our planet's satellite is often not possible for days (and nights). How fine it is to have a watch or clock to show what phase the moon is momentarily in above the blanket of clouds.

A moon phase is usually displayed with the help of a dark blue disk that is rotated by the watch's dial train. Two large, usually gold-colored circles that are printed or polished and located across from each other alternately show themselves in a cutaway in the dial, although only a portion of the circles representing the moon is visible due to the shape of the cutaway in the dial. The moon disk revolves entirely around over a period of fifty-nine days, not quite two months, in most watches with moon phase displays. This is close to the period of time known in watchmaking as the lunar month, which is somewhat longer than 29.5 days (29 days, 12 hours, 44 minutes, and 2.9 seconds to be exact) and is officially termed the synodic moon.

It is easy to see that great imprecision can arise by using this system, for two synodic lunar months are not exactly 59 days long, but rather 59 days, 1 hour, 28 minutes, and 5.8 seconds. A deviation created in this manner can mean a departure of one whole day within a two-and-a-half-year period. The Lange 1 Moon Phase, with its clearly more complicated gear train, shows a deviation of one day in relation to the current cycle of the moon only after a period of 122.6 years.

LANGE 1 Moon Phase

Manufactured since	2002
Movement	mechanical manually wound movement, Caliber L901.5, diameter 30.4 mm, height 5.9 mm, 54 jewels, 9 in screw-mounted gold chatons, glucydur screw balance, Nivarox balance spring with special terminal curve, swan-neck fine adjustment and patented beat regulation, twin serially operating spring barrels, 21,600 vph, power reserve 72 h; Glashütte three-quarter plate, bridges decorated with Glashütte ribbing
Functions	hours, minutes, subsidiary seconds, outsize date (patented); stop-seconds, power reserve indication, moon phase display with continuous drive
Case	diameter 38.5 mm, height 10.4 mm, sapphire crystal, sapphire crystal case back secured with six screws, corrector button for date display on case edge

continuously from the hour wheel, the slowest moving wheel of the gear train and responsible for moving the hour hand. Contrary to the oft-used technical solution in which a round disk displaying moon symbols is moved forward once or twice a day by a cam, Lange Caliber L901.5's moon disk turns continuously.

It is this constant rotation that enables the watch to clearly show more precise results all the time, for a moon phase gear working in this fashion has far less room for error than other mechanisms.

Little Safety Clutch

It is a known fact that one toothed wheel of a gear train can't be in motion without moving the rest. For this reason Lange's designers added a clutch wheel to the gear train for the moon phase—like the one found in dial trains of modern watches so that the hands can be set while the movement is running.

In Caliber L901.5 this clutch wheel, made of beryllium bronze, also has a rotating steel pinion that supports it with two spring-loaded crossings turning with its stem.

When a moon phase gear is running normally, both wheel and pinion turn jointly upon the shoulder screw. If the moon phase display needs to be corrected, the correction lever located above the recessed button on the case engages the pinion's teeth rotating between the wheel's crossings, thus directly connecting it to the moon disk's gear rim.

Moon by Day, Luminous Hands at Night

When the moon is visible in the watch's little firmament, it can only be seen with the aid of outside light. Not so for the hour, minute, and power reserve hands inlaid with a glow-in-the-dark substance that, together with the luminous markers on the dial, can be clearly seen even at night.

The dial is captivating in its typical Lange clarity. The design of the power reserve display is especially clever. Since the designers decided not to put a full scale along the power reserve indication, the empty space that was created on the dial sets off the moon phase display, main dial, and outsize date beautifully. The outsize date, by the way, can be set by the button on the upper left side of the case edge. ■

The Lange 1 Moon Phase in a red gold case.

Lange 1 Time Zone

Two Days, Two Nights

In 2005 A. Lange & Söhne added to its collection with a fully new and unusual world time watch. The Lange 1 Time Zone model features a set of two dials, two pairs of hands, and dual day/night indications.

A. Lange & Söhne has proved again and again in the last few years that one does not have to invent something completely new to create something special. The company has done this interestingly, however, and with impressive inventions showing how to turn horological ordinariness such as date displays into technical jewels. At the same time, it has also shown how to turn highly complicated watch technology with relatively little practical use, such as a split-seconds chronograph, for example, into a design with no less lavish mechanics, making it an exceptionally practical instrument watch, such as the Lange Double Split.

This is also the case with the Lange 1 Time Zone, introduced in 2005 and inspired by the increasing globalization currently en vogue and the generally increasing desire of the public at large to travel, seeing that it needs "a watch with the display of several time zones." This is done in Lange's own way, in the tradition of this firm, with technical solutions that are the result of a manner of working that is free of conventions. In short, a new model made in A. Lange & Söhne's reliable manner.

Reliable can also describe the movement that forms the base of the new watch. Caliber L03011.1, comprising German silver plates and bridges as part of a total of 417 individual components, is based upon the movement of the original Lange 1 introduced in 1994, which with its large power reserve and striking outsize date became the hallmark of the brand. This base caliber is unmistakably present in the Lange 1 Time Zone.

Above: the complicated mechanism for the reference city ring and the combined time zone and day/night indications (light brown). When the button at 8 o'clock is pushed, the city ring turns 15°. At the same time, the dial train of time zone display (below right on the illustration) is turned ahead an hour, taking its day/night display (left of it) with it.

The packaging and certificates of the Lange 1 Time Zone in a three-piece set on the occasion of its introduction in July 2005.

Lange 1 Time Zone

Lange 1 Time Zone

The Lange 1 Time Zone in a platinum case. The explosion illustration below right on page 57 shows that the three large wheels above the three-quarter plate serve to transmit the energy from the base movement to the separately settable dial train.

As expected, this new development disposes of two dials and pairs of hands that can be set independently of each other as well as a rotating city ring featuring the names of representative locations for the twenty-four time zones of the earth.

Additionally, essential day/night indications for each of the time zones are present whose tiny arrow hands always run in a synchronized manner. The watch's wearer has the choice of which of the displays is home time and which is local time while traveling from all of the time zones shown.

Normally, when the wearer flies to the desired destination, the ring featuring the twenty-four printed reference city time zone names is set using the button located at 8 o'clock so that the desired reference can be seen on the little dial located at 5 o'clock. The hour hand of the sub-

Of Time Zones and Zonal Time

Our earth is officially divided into twenty-four time zones that start at the prime meridian and increase in an easterly direction. This fact is due to the west-east rotation of the earth, which moves in a counterclockwise direction as seen from the North Pole and in a clockwise direction as seen from the South Pole. A world time watch can be set for any of these time zones. The zonal time of the part of the earth for which the watch was set is then displayed.

From Donkeys to UTC

In times where greater distances could only be covered very slowly, the differences between local times didn't play such a key role. There are areas of South America in which mountain villages are located about fourteen kilometers away from each other. This is about the distance that a heavily laden donkey could cover over this rough terrain in one day—also representing a type of time zone measurement, if you will.

In the nineteenth century, every region in Europe still had its own local time. Only the advent of train travel around the middle of the nineteenth century, a mode of transportation with which one could all of a sudden travel incredible distances, forced mankind to finally end this time chaos in order to be able to create reliable timetables for the new machines.

One finds in antiquity representations of the earth with surfaces divided by a net of parallels. The prime meridian was something that Ptolemy had already set down when sketching his map of the Canary Islands region (located today at about 15° longitude west). Later cartographers placed the prime meridian in Rome, Copenhagen, Paris, and St. Petersburg before it was finally set in London. Since the earth revolves around its own axis and every meridian is a line between the North and South poles, the placement of the prime meridian is principally all the same and had more to do with politics than anything else.

Confusing the terms time zone and zonal time can lead to misunderstandings in itself, so it's not hard to imagine all the difficulties there were in introducing these concepts. The problems had already begun with the disagreement between perpetual rivals France and Great Britain regarding the placement of the prime meridian, which one party naturally wanted in London while the other of course desired for it to be in Paris.

Because the Royal Observatory—located in London's suburb of Greenwich—had published a nautical almanac in 1767, the British were convinced that the prime meridian should run through the English capital to form the decisive base line for worldwide timekeeping. From then on Greenwich Mean Time, or GMT for short, has been the accepted norm.

Countries with large eastern and western borders such as Russia and the United States, where at some points there were more than 300 local times, naturally suffered the most under the practice of so many different local times. For this reason, in 1875 the American Meteorological Society began the initiative to create a uniform system of time that would be valid throughout the world. It was officially decided in 1884 to divide the earth into twenty-four time

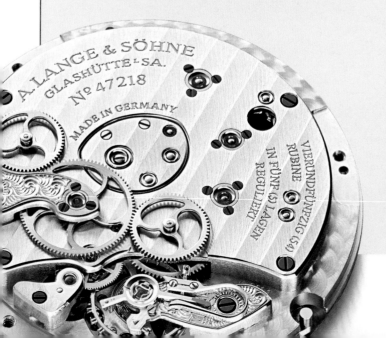

Lange 1 Time Zone

sidiary dial printed with Arabic numerals (at the spot where the subsidiary seconds would normally be placed on the Lange 1) goes forward one hour each time the button is pressed, as does the day/night indication.

During a short stay in a foreign country the setting would certainly be kept as described above. The large dial, to which the date display is coupled, remains all the while on home time. Should one be staying in Karachi or Caracas for a longer amount of time, it might be more comfortable to switch these settings. This can easily be done by activating the correction button at 8 o'clock, and holding it down while setting the hands of the large dial using the crown and holding it pressed while changing the hands of the larger dial by turning the crown in a counterclockwise fashion. ∎

LANGE 1 TIME ZONE

Manufactured since	2005
Movement	mechanical with manual winding, Caliber L 031.1, diameter 34.1 mm, height 6.65 mm, 54 jewels, 4 in screw-mounted gold chatons, glucydur screw balance, Nivarox balance spring with special terminal curve, swan-neck fine adjustment and patented beat regulation, 21,600 vph, power reserve 72 h; Glashütte three-quarter plate, bridges with Glashütte ribbing
Functions	hours, minutes, subsidiary seconds; second time zone; power reserve display; outsize date; home time with day/night indication, local time (hours, minutes) with day/night indication and reference city ring, settable via button
Case	diameter 41.9 mm, height 11 mm, sapphire crystal, sapphire crystal case back secured with six screws; two correction buttons for time zone on edge of case

zones. From one time zone to the next there was thus one hour of time difference, increasing from west to east and using the prime meridian as reference.

Little by little all countries joined the system. Even France finally accepted London as the home of the prime meridian.

Since the day has 24 hours during which the earth revolves once around, making a complete 360° circle, one time zone measures 15° of longitude (360° divided by 24 = 15°). Theoretically.

In practice, however, there are many more time zones for both political and geographical reasons. Some countries would like to have half an hour difference to the next time zone. And in some gigantic countries such as China there is only one valid time zone. Furthermore, many countries do not use daylight savings time, which would, for example, be meaningless in countries close to the equator where the light and dark phases of the day are almost always of the same length.

Central European Time (CET) was introduced in Germany, including Glashütte, in 1893 and corresponds to Greenwich Mean Time plus one hour. It is valid from the west coast of Spain all the way to the eastern border of Poland, thus making it the valid form of time for almost thirty-five lines of longitude.

The term UTC (Universal Time Coordinated) is used more often today than the term GMT and was introduced in 1972. UTC, based on atomic clocks and spread by shortwave transmitters and satellites, is oriented on the prime meridian, but instead of making Anglo time displays such as 5 a.m. for morning and 5 p.m. for afternoon, it uses the more precise military display of time expressed in 24 hours. A four-digit way of writing the time is used that is retained when saying the time out loud. For example, 1:00 a.m. is actually UTC 0100 and is pronounced, "zero one hundred."

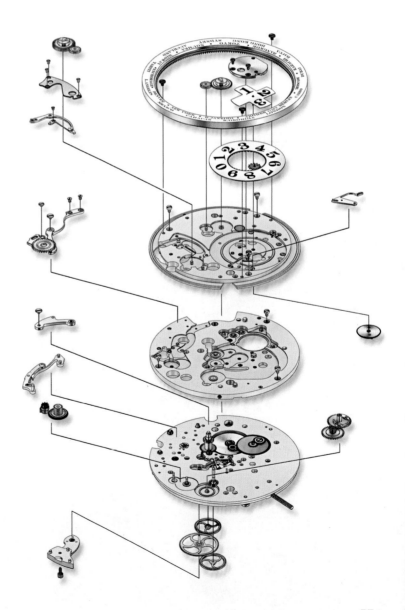

Lange 1 Tourbillon

Complicated Sibling

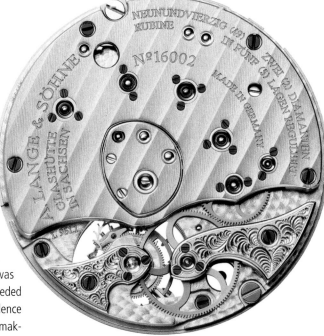

Obvious on Caliber L961.1 is the large transmission wheel connecting the gear train with the tourbillon cage.

It was fairly clear for most friends and connoisseurs of the brand A. Lange & Söhne that after the quick sale of the entire edition of the tourbillon Pour le Mérite, the manufacture would produce a new watch outfitted with this device for cheating gravity in the near future. It was also clear that this watch would have a completely different character than the tourbillon model belonging to the first edition of Lange watches that debuted on October 24, 1994.

Pour le Mérite, sold out in the blink of an eye, was tangible proof of horological competence, proof needed as a signal to the world of watch's top brands. Evidence of the company's carefully maintained art of watchmaking was something that Lange could not produce during its period of expropriation in the GDR era, and thus it was necessary for the company to furnish a new exemplar in order to find its own identity in reunified Germany.

On the other hand, that highly complicated showpiece of the first post-World War II Lange collection, actually outfitted with classic characteristics of pocket watch technology, functioned as a clear message to the traditionalists among connoisseurs of the company (and back then that was probably all of them). The new tourbillon needed to achieve something else.

A Face with a Mind of its Own

The Lange 1 Tourbillon, introduced at the World Fair for Watch and Jewelry in Basel 2000, was certainly a watch to polarize the public, as was previously the case with the original Lange 1 model in the mid-1990s. The designers and watchmakers at A. Lange & Söhne had once again set out on an unusual path.

The position of the escapement alone—that group of parts revolving along with the tourbillon cage—within the movement was unusual. While this is usually located in other watches on the vertical axis between 6 and 12 o'clock, the tourbillon of Caliber 961.1 was brought out of this symmetry by a long bridge looking for all the world like a comet's tail, actually creating its own new type of harmony. Together with the outsize date located on the upper edge of the dial, the carriage tirelessly making its revolutions once a minute built a perfect visual counterweight to the dominant, off-center positioning of the main dial's scale ring.

Certainly one voice or another criticized the large, mirror-polished steel surface of the bridge, so obviously a conscious part of the silver dial. Unbounded, however, is the praise directed at the airy, completely carefree appearance of the carriage's construction. Even the cock, cut down to the boundaries of stability on the front, and

Lange 1 Tourbillon

A special characteristic of the Lange 1 Tourbillon is the steel, mirror-polished cock, whose tourbillon revolves underneath a diamond endstone. The watch, manufactured from 2000 to 2002, was outfitted with a manually wound movement, Caliber L961.1, with a diameter of 30.4 mm and a height of 5.9 mm. It contains forty-nine jewels, nine of which are found in screw-mounted gold chatons; the tourbillon cage bears two diamond endstones. The movement runs at a frequency of 21,600 vph, and thanks to the serially operating spring barrels, like Lange 1's base movement, it also has a power reserve of 72 hours. The case has a diameter of 38.5 mm and is outfitted with sapphire crystals, even in the case back secured by six screws. One hundred fifty pieces in platinum and 250 in red gold were manufactured.

the cage, highly visible from the back of the movement, garnered unanimous approval in horological and consumer circles. The 150 pieces in platinum and 250 in red gold were each also sold out faster than you can say "tourbillon."

A Diamond-Studded Lightweight

The impression of tender fragility that the cage leaves is underscored by a specialty last seen on the best class 1A vintage pocket watches made by Lange: the crystal-clear diamond endstones limiting the upward swing of the carriage pivots.

The idea of lightness transmitted by the sight of the balance busily swinging to-and-fro and the simultaneously quick, slightly jerky rotating motion of the tourbillon cage does indeed have a solid background. The entire tourbillon subgroup and the miniscule cage that holds it comprise seventy-two parts including the balance, balance spring, pallets, escape wheel, and tiny plates and only weigh a total of 0.234 grams!

The diamond endstones are, just like seven of the forty-nine jewels used in the movement, bedded in so-called chatons—gold jewel beds with polished grooves that are secured to the movement's plates and bridges by hand-tempered blue steel screws. Since there is no normal balance cock in this movement (although those found in every other Lange movement are decorated with artful hand engravings making each and every one of them unique), the engravers have decorated the backward cocks supporting Caliber L 961.1's large intermediary wheel and carriage. The latter transmits energy supplied by the famous twin spring barrels through the movement to the carriage containing the balance, pallets, and escape wheel.

The Lange 1 Tourbillon successfully combines the tried-and-true technology of its movement with the filigreed delicacy of its simple yet ingenious complication, thus joining very traditional watchmaking with the unusual design of a modern wristwatch. ■

Grand Lange 1 Luna Mundi

Moon over Buenos Aires

The moon phase's display mechanism: visible is the connection to the hour wheel, at left the lever for quick-setting.

The Big Dipper and Southern Cross constellations embellish the dial of these two moon phase watches.

If one observes the night sky in a country south of the equator, not only do the constellations look different, but the phases of the waxing and waning moon also look different than they do in the Northern Hemisphere—they are in fact mirror images. Logically, only the full moon and new moon are the same in both hemispheres.

Until now watch connoisseurs in Buenos Aires, Auckland, and Jakarta had to be satisfied with moon phase watches that correctly displayed the status of the earth's satellite, but which illustrated the wrong direction of waxing and waning for the Southern Hemisphere, basically backwards for people living there. A. Lange & Söhne decided to occupy itself with this problem and presented a set of new models in the spring of 2003. Under the collection name Luna Mundi ("World Moon" in Latin), these two watches comprise a set illustrating the different appearances of the moon for its owner in an impressive way.

The timepieces, which, by the way, could only be purchased together, differ from each other in the direction of their rotating moon disks and in the design of the case and the dial crafted in solid silver. By leaving out the numeral 12, the hour circle of white gold model Luna Mundi Ursa Major for the northern half of the earth is embellished with a stylized illustration of the Big Dipper constellation. The Lange 1 Luna Mundi Southern Cross, on the other hand, is housed in a red gold case. Hour numerals, date window frame, and the illustration of the Southern Cross constellation located where the 12 is usually found correspondingly shine in a warm reddish hue.

Both constellations are also depicted on the corresponding display disks with the moon symbols rotating underneath the subsidiary seconds dial, partially visible in a classic semicircle cutaway in the dial. The moon of the Northern Hemisphere is visible on its clockwise rotating disk at the upper edge of the window. The southern moon, on the other hand, appears at the lower edge of the cutaway as a narrow crescent and moves in a counterclockwise direction, slowly waxing and waning. The separate scale showing the age of the moon must be read correspondingly either from the left or the right.

Technically speaking, the manufacture of a watch whose moon phases appear analogue to those of the southern sky is so simple that one must actually wonder why such watches were not made previous to this. It is simply the rotation direction of the gear train in the control mechanism for the disk featuring the moon symbols for the Northern Hemisphere that needs to be changed. For this reason, the Southern Cross's movement (Caliber L901.7) only had four components more than its counterpart—it comprises 402 parts while Caliber L901.8 of the Ursa Major has 398.

The globally usable watch duo Luna Mundi was only manufactured in an edition of 101 sets and delivered in a precious wood case that also contained two no less luxurious leather travel cases for the timepieces. ■

Grand Lange 1

The moon phase display of Ursa Major rotates clockwise.

The watches of the Luna Mundi collection, manufactured from 2003 until 2004, differ in their case materials and the rotational direction of their moon phase displays. The manufacture movement Caliber L901.8 shows the moon as it appears in the Northern Hemisphere, for which reason this model was named for the Big Dipper constellation (Ursa Major). In Caliber L901.7, also known as the Southern Cross, the moon phase practically rotates in the opposite direction. Both movements have a diameter of 30.4 mm and are 5.9 mm high. They contain fifty-four jewels, nine of which are embedded in screw-mounted gold chatons, and they are outfitted with glucydur screw balances and swan-neck fine adjustments. The cases have a diameter of 41.9 mm and, depending on the model, were available in cool white gold or warm red gold.

Saxonia

Saxon Beauty

This delicate manual-wind timepiece from the debut collection of 1994 represented a discreet counterpoint to the more technically defined leading model, Lange 1. Its special charm is to be found in the combination of a few rather inconspicuous details.

The Saxonia is available in platinum or yellow gold. Diameter 33.9 mm, height 9.1 mm; sapphire crystal; silver dial; steel hands; gold crown; case back available with sapphire crystal or in solid gold or platinum.

Saxonia—the name is "a patriotic declaration of the Lange family to its homeland." Whatever feelings may be expressed in both this line quoted from A. Lange & Söhne's catalogue and this timepiece, the Saxonia model is nothing other than another example of the timelessness of classic elegance made in Glashütte, which allows all fashion trends to just flow past.

Just as when looking at people, the observer of this watch will also ask himself or herself where the attractiveness lies. What is it that holds your gaze, practically taking the eye prisoner? And here, as there, you may not want to know the answer sometimes—you are afraid the fascination of the image could lose its magic effect if you study the details too closely.

With its unusual off-center dial arrangement, the Lange 1 model ensconced itself in the position of brand icon during the presentation of A. Lange & Söhne's first wristwatches in 1994. On the flip side, however, it was clear to everyone that this design could polarize its audience, and thus the company created the Saxonia as a peace offering to the more conservative among the brand's admirers.

The name "Saxonia" is one that is present to this day in the memories of all those who went to school in the little Erzgebirge city Glashütte. For them it is a symbol of human togetherness and collegiality among Glashütte watchmakers. In the fall of 1885, students of the German School of Watchmaking (DUS) created a club by the name of Saxonia whose aim was to promote personal contact among the apprentices in an undogmatic manner. This club developed so unexpectedly well that in 1904 its continuation, Altherrenverband der Schülervereinigung Saxonia an der Deutschen Uhrmacherschule (Veteran's Organization of the Student Club Saxonia at the German School of Watchmaking) was formed, and even had its own club newspaper. The former members met twice a year, and local branches were created in larger cities, within which members met to talk shop and dedicate a bit of time to the watchmaker students of the following generations.

Less is Sometimes More

Since at A. Lange & Söhne even a three-handed watch with an outsize date display is not just developed on the spur of the moment, more than thirty design sketches were needed before the company decided upon the final appearance of this timepiece.

The Saxonia model is reduced to the minimum—at least it would seem that way. A simple, undecorated

Saxonia

Saxonia

case made of platinum, white gold, or yellow gold with powerful lugs and a wide bezel; slim lance-shaped hands made of gold or silver, and a very delicate minute ring located on the edge of the solid silver dial—so far nothing unusual.

But then the eye is drawn to the instantly likeable outsize date in its double gold-framed aperture. Almost too dominant, it finds its counterpart in the overly dimensioned "subsidiary" seconds dial.

This subdial takes up almost half of the full dial's diameter for itself and demands attention, aided and abetted by the slim second hand and its conspicuous rhombus-shaped tip.

Almost as if it were the connector between these two eye-grabbing displays, the name A. Lange & Söhne, which is placed on the upper half of the dial on almost all other Lange watches, has taken a neutral position, along the imaginary line between the nine and the three. Thus, the watch's designers have reached their goal of having the Saxonia, despite the dominance of the outsize date display, taken in as a whole.

The rectangular button for quick-setting the date, with its satin finish and polished, beveled edges, follows the unspoken order of "no one gets the spotlight" and is so well integrated into the case that it seems to be practically a part of it.

Lange Caliber L941.3:
It has a diameter of 25.6 mm and a height of 4.95 mm. Three-quarter plate made of German silver with Glashütte ribbing; four screw-mounted gold chatons; hand-engraved balance cock; swan-neck fine adjustment; stop-seconds. The movement is regulated in five positions and has a glucydur screw balance that runs at 21,600 vibrations per hour.

Real Beauty Comes from Within

Although the real face of the Saxonia model is somewhat harder to recognize for fans of fine watches, the timepiece shows itself in all of its effusive beauty when viewed through its sapphire crystal case back.

Lange Caliber 941.3 is a true eye catcher. Four gold chatons, whose jewels serve as upper bearings (for watchmakers, the back side of the movement is always the "top" side) for the core, minute wheel, third wheel, and fourth wheel, are secured by deep blue steel screws to the three-quarter plate made of matte German silver. The base plate, just like the hand-engraved balance cock, is also fastened with fully finished steel screws that receive their dark blue color from a heat treatment after hardening.

The three-quarter plate is decorated with Glashütte ribbing, and engravings flushed out with gold are abundant on its surface. Except for "made in Germany," all of the other words are in German. The engraved texts hardly need explaining for the watch enthusiast, though. The regulation in five positions mentioned (IN FÜNF LAGEN REGULIERT) means that the rate of the watch was tested and regulated in the positions "dial side up," "dial side down," the so-called hanging position "crown down," "crown up," and "crown to the left." "Crown to the left" corresponds to a watch worn on the left wrist of a wearer sitting with his arm resting on a table.

The escape wheel's endpiece, secured by screws from the bottom, is finely polished, as are the screw of the angle lever that keeps the winding stem in place and the rounded ends of the gear train and core's pivots. The spring of the swan-neck fine adjustment, the so-called balance spring stud carrier, and the rounded end, visible from above, of the triangular stud are polished as well. The triangular stud, although it could just as well be square or pentagonal, is shaped this way so that it can return to exactly the same position within the balance spring stud carrier as it had before should the balance ever need to be taken apart.

The balance spring stud, a little entity made out of brass or German silver with a miniscule hole drilled through it containing the last loop of the balance spring, is generally of a cylindrical shape in most simpler watch

Saxonia

movements. Thus, it can rotate in its drilled hole and have a different position should it ever be taken out. This would, however, cause a different (even if very slight) course for the outermost balance spring loop as a result so that it would no longer fit exactly through the regulator index. That would, furthermore, have an influence on the watch's rate.

Delicate in Every Way

While the platinum version with its rhodium-plated dial and shiny rhodium-plated hands represents sheer understatement, it is the gold models that are best suited to embellishment with jewels on the bezel. Less than 34 mm in diameter, this Saxon beauty is more qualified for gracing feminine wrists in the European and American markets, while in Asia it embodies the perfectly fine and noble gentlemen's watch. ■

A special detail of the Saxonia is the button for quick-setting the outsize date. To the right of that another special element becomes recognizable: the finely cut case lugs.

SAXONIA

Manufactured since	1994
Movement	mechanical manually wound movement, Caliber L 941.3, diameter 25.6 mm, height 4.95 mm, 30 jewels, 4 in screw-mounted gold chatons, glucydur screw balance, Nivarox balance spring with swan-neck fine adjustment, patented beat adjustment via micrometer screw, 21,600 vph, power reserve 42 h; Glashütte three-quarter plate, bridges decorated with Glashütte ribbing
Functions	hours, minutes, subsidiary seconds, outsize date (patented)
Case	diameter 33.9 mm, height 9.1 mm, sapphire crystal, sapphire crystal case back secured with six screws; corrector button for date display on edge of case

Arkade

Bigger Inside than Out

The rows of arcades at Dresden's castle stood model for the designers of Lange's Arkade model, and even the double-digit date window is inspired by one of Saxony's cultural assets: the famed digital Five-Minute Clock located in the Semper Opera. While ladies might certainly enjoy the stylishly jeweled Arkade models, gentlemen are probably more occupied with how in the world the mechanism for the outsize date display even fits into the slim case …

The double-digit outsize date was the star of the show during the first ad campaign for modern-era Lange watches, and as an illustrative motif the agency chose not to show the collection's flagship model, Lange 1, but rather the delicate women's watch, Arkade. Its small shaped movement was put on display with its gigantic date ring. The numerals printed on this huge ring are the same size as the date shown in the Arkade's prominent double aperture. The reader was then asked how this large date display found room in such a small watch, though he or she immediately received an answer worthy of an oracle, "… only Lange knows."

Even if the veil of mystery was soon lifted in the manufacture's catalogue as well as in numerous newspaper and magazine articles with remarkable transparency and technical details, the fascination for this technical design trick remained, and with it, a great deal of respect for the sole women's watch of 1994's debut collection.

Historical Models

A. Lange & Söhne is not only a traditional brand; today's incarnation of the company really uses its history to advantage. Discreetly and in a restrained manner, it is in no way blatant or vociferous, as some other brands are wont to be. Not only does Lange refer to its more than 150 years of history, but also to the German state of Saxony, that region surrounding the state capital Dresden—one of Germany's most beautiful cities—and the little city of Glashütte nestled in the Erzgebirge mountain range.

The impressively clean shape of the Arkade case is, for example, modeled upon the arcades of the castle in Dresden with its arcs, "a symbolic portrayal of Saxony's legendary glory" (quote from Lange's catalogue).

The "legendary glory" of the manufacture founded in 1845, whose watches were already being constructed in the era of the last German chancellor, is simultaneously a synonym for beauty and perfection in watchmaking, known as prestige objects *par excellence*. This is something that the Arkade lived up to easily, the legendary certainly being highlighted by the models of 1994, with presentation in simple gold cases, of which, if anything, only the bezel was set with diamonds. In the meantime these timepieces have been succeeded by newer models pleasing the eye with a veritable fireworks display of brilliant-cut diamonds and other gems. However, the purchase of one of these watches does leave an obvious hole in one's bank account, the extent of which is hidden behind the words "price on request" in Lange's catalogue.

Digital Watch from the Nineteenth Century

The large digital display of the date has, alongside the technical innovation and practical use it represents, a tie to the history of the manufacture, the city of Dresden, and, last but not least, to the familial background of the founder, Ferdinand Adolph Lange.

The date display, today a definitive Lange trademark, is supposed to remind one of the first clock featuring a numerical display, made in 1841 for the Dresden Semper Opera House, the so-called Five-Minute Clock. The Lange outsize date appears in a double aperture framed in gold applied to a dial made of solid silver.

Lange & Söhne's outsize date is an extremely practical innovation, enjoyed not only by people in need of reading glasses. This was also the first time that a date display on a

Lange Arkade: yellow gold case, 29 x 22.2 x 8.4 mm; sapphire crystal; silver dial; gold hands; gold crown; case back available in either solid metal or sapphire crystal; quick-set date via button.

Arkade

watch was used as a stylistic element, a dominant detail in the design of the dial. A date display generally lives its little life like Cinderella—unobtrusively in a small, usually rectangular or square aperture in an out-of-the-way corner on the dial. If it receives a subsidiary dial, hand, and attractive appearance, it will usually detract from the practical side of the display, often proving hard to read.

The Lange watches, unusual enough to start with, were made unmistakable by the invention, showing innovative competence in technology and a practical side in comparison to other watches while remaining within the confines of the brand's tradition. The date display can be quickly adjusted by activating the recessed button on the right side of the case edge. The stylus necessary to complete this procedure is delivered with the watch.

But it's not only the outsize date that made the Arkade a very special women's watch. Its 18-karat gold case, produced to a large degree by hand, with the dimensions 29 x 22.2 x 8.4 mm, shelters a treasure in the art of watchmaking—shaped Caliber L911.4, designed and built in the best of the manufactory's tradition.

The movement harmonizes wonderfully with its case, as both are identical in shape. The movement, measuring 25.6 x 17.6 mm in diameter and 4.95 mm in height, can be admired through a sapphire crystal case back, although the owner of an Arkade can also choose to purchase the watch with a solid gold case back instead. The visual pleasure of the movement is intensified by the three-quarter plate and the escape wheel bridge decorated with Glashütte ribbing, the three screw-mounted gold chatons, and an engraved balance cock.

An early digital clock: This is a model (ca. 25 cm high) of the Five-Minute Clock located in Dresden's famous Semper Opera. On the left side, the clock shows the hours in Roman numerals, and on the right side the minutes are displayed in intervals of five minutes in Arabic numerals. The small analogue clock above the pendulum serves only as a control.

The movement is regulated in five positions and its industrious glucydur screw balance swings busily at 21,600 vibrations per hour. The effective length of the Nivarox balance spring can be changed by turning the micrometer screw of the swan-neck fine adjustment should this exclusive movement ever begin to lose or gain time. ∎

ARKADE

Manufactured since	1994
Movement	mechanical manually wound movement, shaped Lange Caliber L 911.4, 25.6 x 17.6 mm, height 4.95 mm, 30 jewels, 3 in screw-mounted gold chatons, glucydur screw balance, Nivarox balance spring with swan-neck fine adjustment, patented beat adjustment via micrometer screw, stop-seconds, 21,600 vph, power reserve 42 h; Glashütte three-quarter plate, bridges decorated with Glashütte ribbing
Functions	hours, minutes, subsidiary seconds, outsize date (patented)
Case	29 x 22.2 mm, height 8.4 mm, sapphire crystal, sapphire crystal case back secured with four screws; correction button for date display on edge of case

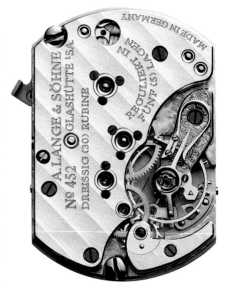

Arkade

Disk Date

The large date by A. Lange & Söhne is based on a completely new design, the realization of which was supported by a patent that Jaeger-LeCoultre filed in the 1930s but never used.

In normal date displays, a gear train is placed on the dial side of the base plate next to some levers and a click, depending on the type of construction chosen. One of these wheels interacts with the hour wheel and is driven by it. The so-called date change gear moves a ring located around the outer edge of the movement printed with the numerals from one to thirty-one either constantly for several hours at a time or with a sudden shift at midnight. The previously mentioned click serves to keep the date ring in a resting position after the date has changed so that the current numerals remain exactly underneath the window on the dial even when the watch is in motion.

When designing A. Lange & Söhne's outsize date, the engineers realized right from the beginning that they had to distance themselves from the usual large ring with numerals, for even the slightest enlargement of the numbers would have made a gigantic ring necessary, which

The dial side of Caliber L911.4 with its overly dimensioned date display. The single digit disk, whose edge just barely clears the fourth wheel arbor, is easily recognized here.

may have worked in a man's watch, but certainly not in a timepiece like the Arkade.

It has always been the practice of watchmakers to arrange parts that would otherwise make the diameter of a movement too large on top of each other. Following this old principle, Lange's chief designer Helmut Geyer and his coworkers developed a date display with two rotating

Top: The Lange outsize date contains disks for both single and double digits that are combined to display the correct date. The distance between the disks, not touching although they are stacked, is 0.15 mm. The eye registers the complete date as if it were on one level, an illusion necessitating very exact positioning of all participating components.

The program wheel for the single digit is moved by an adjoining wheel (not visible), which in turn moves the impulse wheel (blue). The program wheel is "missing" three teeth, so that, although it rotates every day, the digit does not change one day each month (from the 31st to the 1st).

69

Arkade

The Lange outsize date's entire mechanism comprises 66 tiny precision parts.

disks located one above the other, one in the shape of a ring as large as the case of Lange's smallest timepiece would allow and the other in the shape of a cross.

In regular digital date display designs containing printed, rotating disks, diameters are limited to the space between the hand arbor and the inside edge of the case. From time to time, these disks extend beyond the edge of the movement, giving those mechanisms a Mickey Mouse–type shape. The off-center hand arbor of the Lange Arkade is located within the round disk's radius of rotation.

A mechanism to move both of the disks via the hour wheel and some additional wheels necessitated a large amount of constructive creativity, but was technically easy to realize. The decisive difference from a normal date display is the separate cross-shaped numeral disk displaying the first digit as well as the ring-shaped disk, upon which the numerals zero through nine only instead of all thirty-one need to appear, for the second digit. This was the only way possible to enlarge the display field by just about five times for the Arkade model (and three times for the other Lange watches) in a relative comparison to normal date displays. This degree of legibility is otherwise unattainable.

The ring-shaped disk printed with the numerals one through zero rotates daily. The cross-shaped disk on top of it, bearing the numerals one through three and a white space, is only moved every ten days. From the 1st to the 10th of every month, the cross-shaped disk shows its white space on the left side of the date window, and from the 10th its first digit. For months with fewer than 31 days, the date display can be adjusted by activating

Arkade

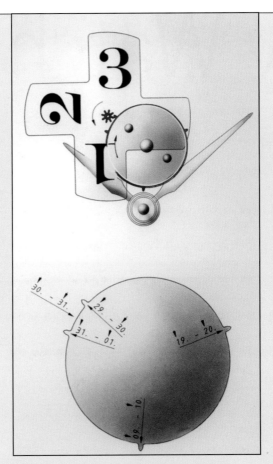

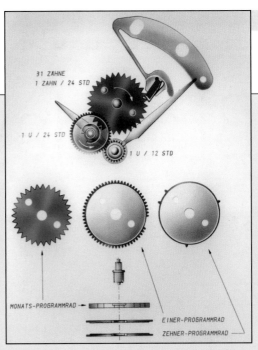

The program and impulse wheels of the date change: The toothed wheel belonging to the hour hand moves the date-change wheel (blue) every 24 hours one tooth (one day) along. The date-change wheel and the program wheels controlling the date display are mounted on the same arbor. A jewel in a notched spring ensures that the date-change wheel remains where it is until the next day.

The cross-shaped disk, only 0.1 mm thick. Its four-toothed program wheel changes the numeral once every ten days, although when the three appears, it moves after only two days.

the recessed button on the case with the stylus that is delivered with the watch.

The designers did, however, have problems with the seven months containing exactly 31 days. Since the disk for the right-hand digit is moved every day, and the cross-shaped disk only every ten days, normally the number 32 would follow the 31st of each month. A technical solution was needed to ensure that the single digit would not change on the first day of a month and thus show "one" two days in a row—on the 31st of the old month and the 1st of the new month. Furthermore, it was necessary to ensure that the cross-shaped disk with the numerals one to three and the white space—contrary to its programming—remained at three for only two days before moving on to the white space. The technical solution, simply put, looked like this: The toothed wheel for changing the single digit is missing three teeth, while the almost "toothless" wheel for the double digit (one for each numeral on the cross-shaped disk) received a fourth tooth.

Absolutely correct positioning is imperative for the components of the date-changing mechanism, for between the cross-shaped disk and the disk below it showing single digits there is only 0.15 mm of space. It's easy to imagine that these two components, only a tenth of a millimeter thick themselves, might chafe each other if they do not rotate in an absolutely true manner.

The Lange date mechanism possesses three of the previously mentioned clicks. These miniscule steel parts, outfitted with even tinier jewels, are there to reduce friction on the toothed wheels for changing the date and the display disks.

Although it was described in a simple manner above, this mechanism is very complex. Its sixty-six individual components are located both above and below a separate plate. Nevertheless, the height difference between both display disks cannot be detected by the naked eye. Especially because the edge of the cross-shaped disk is located exactly underneath the gold mullion of the date window on the dial, the eye is perfectly deceived by it. ∎

1815

The Birthday Watch

If one wanted to be dramatic, one could say, "On February 18, 1815, a new and—as it would become—unique chapter in the history of timekeeping would be written, ending in a manufacture for watches of quality unknown to that point." On the other hand, without the addition of decorative details, this information would simply read: "Ferdinand Adolph Lange's date of birth was February 18, 1815."

Eighteen-fifteen, the year the Glashütte watch industry's founder was born—described with or without embellishment—was reason enough for the contemporary watch manufactory A. Lange & Söhne to name a watch for it 180 years later in 1995: Lange 1815. Since Ferdinand Adolph Lange founded his own company at the age of thirty, it was exactly 150 years ago, on December 7, 1845, that Lange & Comp. was established, the company from which the world-famous manufacture would arise. In Glashütte no one celebrated the 150th anniversary of Lange's birth, for from April 20, 1948, Lange & Söhne was expropriated and turned into a "people's company." That is, until Walter Lange, great-grandson of the founder, once again registered the company on December 7, 1990, reactivating the most important name in German watchmaking, much to the joy of watch fans the world over.

Lange 1815 represented the manufacture's fifth fully new construction since its refounding in 1990, and in the heralding press release, alluding to Wilhelm Busch's playful story of *Max and Moritz*, it was termed "The Fifth Trick ..."

1815

The "Fifth Trick" is Manually Wound

The Lange 1815 model is a classic men's wristwatch with a clean, round case and an impressively clear dial. On the outside, the 1815 offers nothing further than the current time comprising hour, minute, and subsidiary seconds. And it is exactly this watch's reduction to the essential that is so fascinating.

Contrary to the watch's clean outer beauty, the remarkable movement seen through the sapphire crystal case back is effusive in its horological charm. The jewels for the barrel arbor, minute wheel, third wheel, and fourth wheel, bedded in gold chatons, are secured to the three-quarter plate decorated with the Glashütte ribbing that has always been so typical of Lange with deep blue screws. The escape wheel rotates under an endpiece screw-mounted from the inside, and the glucydur balance swings busily underneath a hand-engraved cock upon which the polished spring of the swan-neck fine adjustment sits.

The 1815 model, an eye-catcher perfectly designed with the motto in mind that less is often more, demands daily attention and affection. A hand-wound watch must be provided with the power the mainspring needs in a timely manner. It makes sense to do this in the morning, for during the course of a day a watch often has to remain resistant to uncomfortable positions and shocks and does its best laden with the energy of a fully wound spring and the resulting powerful swinging of the balance. A watch is not only irritating when it stops, it can also cause embarrassing situations.

Regular winding is the rule of thumb for a manually wound watch, and the owner should always remember to do this at the same time every day—especially if he would like to test the rate for several days in a row. For the man who is not quite this disciplined, A. Lange & Söhne has also created a companion to this model that discreetly reminds him when it is time to lavish attention upon it.

1815

Manufactured since	1995
Movement	mechanical manually wound movement, Lange Caliber L 941.1, diameter 25.6 mm, height 3.2 mm, 21 jewels, 4 in screw-mounted gold chatons, glucydur screw balance, Nivarox balance spring with swan-neck fine adjustment, patented beat adjustment via micrometer screw, stop-seconds, 21,600 vph, power reserve 45 h, Glashütte three-quarter plate, bridges decorated with Glashütte ribbing
Functions	hours, minutes, subsidiary seconds
Case	diameter 35.9 mm, height 7.5 mm, sapphire crystal, sapphire crystal case back secured with six screws

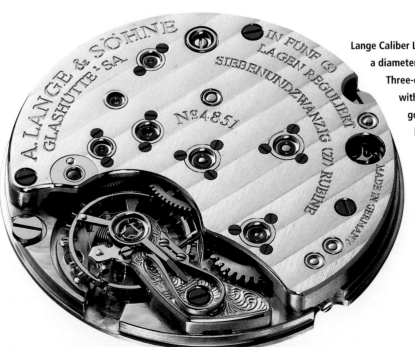

Lange Caliber L942.1 of the 1815 UP and DOWN: It has a diameter of 25.6 mm and is 3.7 mm high. Three-quarter plate made of German silver with Glashütte ribbing; 6 screw-mounted gold chatons; hand engraving of the balance cock; swan-neck fine adjustment; stop-seconds, integrated power reserve mechanism. The movement is regulated in five positions and its glucydur screw balance works at a frequency of 21,600 vph.

1815 UP and DOWN

1815 UP and DOWN

Manufactured since	1997
Movement	mechanical manually wound movement, Lange Caliber L 942.1, diameter 25.6 mm, height 3.7 mm, 27 jewels, 6 in screw-mounted gold chatons, glucydur screw balance, Nivarox balance spring with swan-neck fine adjustment, patented beat adjustment via micrometer screw, stop-seconds, 21,600 vph, power reserve 45 h; Glashütte three-quarter plate, bridges decorated with Glashütte ribbing
Functions	hours, minutes, subsidiary seconds, power reserve display
Case	diameter 35.9 mm, height 7.9 mm, sapphire crystal, sapphire crystal case back secured by six screws

1815 UP and DOWN
The Swiss Call it Réserve de Marche ...

A. Lange & Söhne's 1815 UP and DOWN model saw the light of day in 1997 as a sibling of the puritanical Lange 1815 model created two years previously. As similar to the firstborn as a sibling can be, the younger kin still displays several clear differences when observed more closely. One of these differences can be found in the watch's "face," which clearly states without ever uttering a sound whether it is currently filled with energy or whether it needs some attention.

It's actually only the hand on this watch's subsidiary dial located to the left between seven and nine o'clock that goes UP and DOWN (AUF und AB). The device and its display, labeled with abbreviations of the German words *aufgezogen* (wound) and *abgelaufen* (unwound), are more succinctly called in French *réserve de marche*, *Gangreserveanzeige* by German watchmakers, and power reserve indicator by speakers of English (see info box). Basically, all of these fancy terms mean that an additional mechanism—one that is complicated to construct and that uses the aid of a hand to measure the time since the watch was last wound to the time it should next be wound—is hidden within the watch. The word *Gangreserve* is written on the dial of the 1815 UP and DOWN model, measuring the amount of power reserve in hours.

This type of display has its origin in marine chronometers and deck watches: A. Lange & Söhne supplied the imperial navy with such timepieces in the nineteenth century. Absolute precision was so important for them as even the slightest deviation could lead to the navigation instruments causing a so-called indication error of several miles—something that could lead to disaster for both ship and crew, especially when near a coast. At 50° latitude, for example, at the southern coast of England, a one-minute time difference would mean an indication error of just about ten sea miles. The nearer to the equator the ship is, the larger the inaccuracy. Although several clocks were

1815 UP and DOWN

A Reserved Peace of Mind

The Lange 1815 UP and DOWN's power reserve indicator comprises a scale covering about five-sixths of a circle and a blued-steel hand spanning the entire scale in approximately 45 hours. The scale's measuring unit is the hour, and every eighth hour is marked by a numeral. The key area between eight and zero is marked with even smaller divisions.

If the hand reaches this area, it signals to the watch's owner, just as the gas tank display of an automobile, that it's almost time to fill the "tank." He or she should soon wind the watch's mainspring.

A power reserve display, also called an UP and DOWN movement in Glashütte watchmaker jargon, is connected to the watch's "tank" or spring barrel by a complicated gear train. There is also a connection with the wheels of the watch's winding mechanism. A gear leads the display's hand toward the AB on the dial when the mainspring only has a certain amount of energy left to keep the movement running. When the watch is wound, the wheel carrying this hand is pushed in the opposite direction.

What is easy to see when this function is observed closely is that the power reserve display only indirectly gives information about the amount of energy left in the mainspring. It is more a mixture of a revolution counter and a type of countdown counter that measures how many revolutions the spring barrel has made since the mainspring was last fully wound.

Since the watch's wearer is hardly interested in the number of times the spring barrel of his or her watch has revolved since the last winding, the scale shows instead the relation that this has to the amount of time that is left before the watch must once again be wound.

An automobile's mechanical tachometer is another example of this principle. It also counts the revolutions of the car's wheels (which doesn't interest a soul), though the dashboard display shows the speed of the car in mph.

always located on board, it was an absolute necessity to avoid having one of them come to a stop.

Thus, the movement driving Lange's 1815 UP and DOWN model is by no means an imitation of Caliber L941.1, ticking in the gold and platinum cases of older sibling 1815. The 1815 UP and DOWN is outfitted with Caliber L942.2, a movement that, in addition to the exact time and position of the mainspring, also displays some familial similarities, though it was for the most part a brand-new construction. The power reserve device alone comprises forty-six individual parts.

The design used as a model was the planetary gear for which Otto Lange, one of the founder's grandsons, received a German Reich patent in 1940. These gears had to be reduced considerably in size, however. A noteworthy fact is that the Lange 1815 UP and DOWN does not utilize module construction, an industry norm that involves screwing any additional mechanisms on top of the existing movement. The Lange power reserve mechanism is certainly a part of the watch movement and, as such, is fully integrated into it.

Caliber L942.2 has the same diameter as its sibling movement from the 1815 model, L941.1, but at 3.7 mm it is half a millimeter higher and it carries six gold chatons on its three-quarter plate.

1815 Automatic

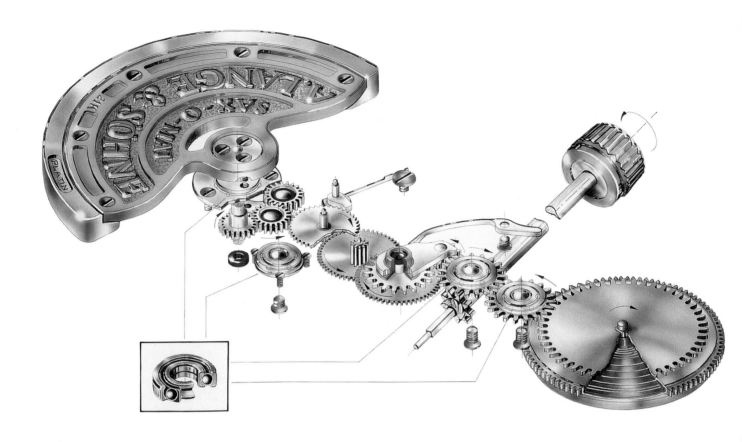

All Wound Up

The circle segment of the SAX-O-MAT automatic rotor is made of 18-karat gold, the ballast weight on its outer rim of 950 platinum. Platinum with its specific weight of 21.45 (gold has a specific weight of "only" 19.3) counts as one of the heaviest metals known and is superbly suited as a "weight strengthener," though it is only utilized in luxury watches due to its high price. Both of these heavy metals "really get the rotor going," actively supported by the clever technology of the rotor bearing, comprising a ball bearing and a ruby.

Automatic rotors running on ball bearings have been an industry standard since the invention of the micro ball bearing by the Swiss watch factory Eterna in 1948. Each miniature ball bearing usually contains five to seven tiny steel balls, just about 0.6 mm in diameter. Weighing only one thousandth of a gram, these miniscule balls are so light that they float in water. Nothing new thus far.

New in a ball bearing rotor is, however, the additional support of a jewel located underneath the automatic bridge, which can also be adjusted there. This jewel bearing stabilizes the lead of the rotor axle, which has an effect on the flat positions of the watch, minimizing the increased effect of jewel friction and guaranteeing a better flat movement of the rotor.

The automatic winding of the SAX-O-MAT, which, by the way, bears the official caliber number L 921.2, was designed very conservatively. A. Lange & Söhne's designers didn't use click wheels, for example, something found in most automatic watches today. Instead Caliber L 921.2 works with a traditional gear simply called "the reverser" in watchmaking jargon. This gear "changes" its rotation direction when that of the rotor changes, winding the movement in both directions. This is necessary because the movement's mainspring can naturally only be wound in one direction.

The reverser invented by A. Lange & Söhne comprises two tiny steel balls located in jewel beds on a movable bar and rotates both together with the gear rim on the rotor stem, and, additionally—alternately—the automatic ratchet wheel.

The fact that the bar and reverser rest upon a second ball bearing, giving them the largest possible amount of space and ease of motion, is new. This ball bearing is exactly in the middle of the bar, secured to it from the bottom. Both of the previously mentioned little steel balls are located at its tip, giving the bar the effect of a sliding gear.

If the rotor is rotating to the right, as seen from the back of the movement (please see illustration), a connection between the rotor's gear rim and one of the reversers is created, and the rotating movement of the rotor is transferred to the automatic winding wheel through this reverser. The second reverser runs in neutral while this is happening. The winding wheel is secured by a click and can only turn in one direction.

1815 Automatic: Twin Sibling Nonstop

In 2004 the 1815 line received an addition to the family. This young one looks quite similar to its sibling born in 1995. Only the printed term "SAX-O-MAT" gracing the space above the subsidiary seconds dial and the words "Zero Reset" located within the railroad *minuterie* allow anyone to recognize the difference. These elements reveal themselves when the back of this attractive timepiece is observed. The regular manually wound movement of the 1815 is not what is found underneath this sapphire crystal case back, but rather the SAX-O-MAT automatic movement introduced in 1997 with its off-center three-quarter rotor made of gold and platinum and patented zero-reset mechanism. This extraordinary timepiece exclusively wears red gold, yellow gold, or platinum and answers to the name 1815 Automatic.

Men—especially those who love a classic—will be standing in line to get a good look at this. ■

If the rotor changes its rotation direction, the connection between the reverser and the winding wheel terminates. The sliding gear rotates on its ball bearing so far that the second reverser makes a connection with the ratchet wheel, and energy is now transferred from the rotor to the first reverser, moving the second reverser and setting the winding wheel in motion.

The winding wheel's motion causes the actual winding of the movement's mainspring to begin. Using the click, energy that originates from the motion of the wearer's arm is stored and supplied to the movement. The winding wheel doesn't directly wind the mainspring—even the energy of the heavy rotor isn't enough to do that. The aid of several wheels and gears is enlisted to achieve this. Two of these wheels also rotate on ball bearings in order to keep friction minimal. And because of this, the rotor rotates with incredible lightness.

1815 AUTOMATIC

Manufactured since	2005
Movement	mechanical automatic movement, Lange Caliber L 921.2 SAX-O-MAT, diameter 30.4 mm, height 3.8 mm, 36 jewels, glucydur screw balance, Nivarox balance spring with swan-neck fine adjustment, patented beat adjustment via micrometer screw, bilaterally winding embossed three-quarter rotor in 21-karat gold and platinum with 4 ball bearings in the reverser gear; 21,600 vph, power reserve 46 h; bridges decorated with Glashütte ribbing
Functions	hours, minutes, subsidiary seconds, stop-seconds mechanism with automatic resetting of the second hand (zero reset function) when the crown is pulled
Case	diameter 37 mm, height 8.2 mm, sapphire crystal, sapphire crystal case back secured by six screws

1815 Moon Phase

New Moon and the Big Gold Dipper

The 1815 Moon Phase was available in either a red gold or platinum case, limited to 250 and 150 pieces respectively. Diameter 35.9 mm; height 7.95 mm; sapphire crystal; silver dial; gold hands; gold crown; sapphire crystal case back.

With the re-creation of the 1815 model to accommodate an additional moon phase display, A. Lange & Söhne not only produced another timepiece for connoisseurs, but once again gave the watch fan a piece of horological eye candy. Unfortunately, the 1815 Moon Phase has long been sold out.

However, its innovative technology deserves a mention in the context of this publication. The window for the moon phase display in conjunction with the subsidiary seconds dial and the large 12 denoting the hour form an almost equilateral triangle. While some of the hour markers are dots, 3, 6, and 9 o'clock are marked by stars. More stars surround the round window in which the moon can be seen in its varying phases and form the constellation of the Big Dipper. For those not so well versed in astronomy, the constellation is outlined with fine lines.

The Moon Phase Display in General ...

This watch commemorates Ferdinand Adolph Lange's second son Emil. Emil Lange directed his deceased father's company together with his older brother Richard at the beginning, and then alone after 1887. Alongside his vocation as an entrepreneur—one that helped A. Lange & Söhne to world fame—Emil Lange was an active supporter of the watchmaking trade and was president of the board of the German School of Watchmaking in Glashütte for many years. Without a doubt Emil Lange would have been pleased by this watch, the precision of whose moon phase gears can stand up in any comparison to the great astronomical clocks produced in the region.

In general the phase of the moon is displayed on timepieces by a dark blue disk that is rotated by the watch's dial train located underneath the dial. Two large, chiefly gold-colored circles that are printed or polished and located across from each other alternately show themselves in a cutaway in the dial. The moon disk revolves completely in 59 days, or not quite two months.

1815 Moon Phase

... and in Particular

It is almost a matter of course to say that the designers at A. Lange & Söhne once again chose an alternate constructive path. Movement designer Annegret Fleischer developed a mechanism whose round disk rotates under the dial over a period of three moon phases, thus in 88 days and just about 15 hours. This disk shows a portion of its surface in the round cutaway of the dial aperture and is moved twice a day by a complicated gear train control in increments that are barely noticeable. This moon phase gear train is driven by the hour wheel (bearing the hour hand).

The main color of the Lange moon phase disk is not dark blue, but gold. And that of the platinum case's moon disk is rhodium-plated, thus silver-colored.

Placed at a distance of 120 degrees from one another, each one occupying one-third of the disk, are three printed black circles. The diameter of these circles correlates to the size of the round aperture in the dial for the moon phase display. Three times during the almost three months it takes for the moon phase disk to completely revolve, the window is entirely filled with the one of the black circles—the new moon.

The New Moon Watch

One could almost say that the Lange moon phase watch is mainly a display of the new moon, for strangely enough "new moon" is the name for that short amount of time when no moon is visible at all. On the disk, the chubby-cheeked satellite, last seen as the thin crescent of the waning moon, seems to have disappeared from the firmament, reappearing a few days later as a paper-thin arc facing the other way.

In Fleischer's design, the moon phase disk rotates clockwise. For this reason, just one day after the aperture is completely filled with black, a very fine gold or silver-colored crescent facing left appears on the edge of the cutaway—a waxing moon.

Further on in the month, more and more of the unprinted part of the moon phase disk revolves into the window until the cutaway is finally filled with it—the full moon. The next of the "new moons" is already lying in wait on the right edge of the window, slowly gaining ground. The fat full moon then languidly revolves into a crescent facing to the right—a waning moon.

The display is so precisely calculated that only after 1,058 years would it show a deviation of one day. Assuming, of course, that the owner keeps winding the watch regularly until that point. ∎

Lange Caliber L941.1 from the normal 1815 model served as the base for the 1815 Moon Phases's movement. Caliber L943.1 has a diameter of 27.5 mm and is 3.85 mm high.

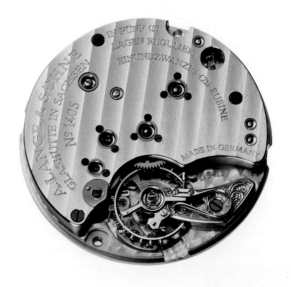

Cabaret

Minority Report

The Cabaret model represents Lange & Söhne's most unusual men's watch, one aspect of which is clearly in the spotlight: its rectangular shape.

Lange's Cabaret: yellow gold case, dimensions 36.3 x 25.5 mm; height 9.1 mm; sapphire crystal; silver dial; gold hands; gold crown; sapphire crystal case back.

And thus the Cabaret model differs from all of the manufacture's other watches simply in its appearance. Even the Arkade, also outfitted with two long parallel edges, bears no similarity to the Cabaret. Thus it remains to point out that Lange's engineers and designers were yet again successful in creating an unusual, completely individual watch model.

The manufacture's catalogue formulates the idea behind this timepiece fairly clearly: "protection of minorities." The "discrimination against rectangular mechanical timepieces by watch connoisseurs" needs to come to an end. Of course, the movement of the previously mentioned Arkade model could have been used in this rectangular case, and that is exactly what most other watch companies would have done. A. Lange & Söhne, however, preferred to create a new movement in order to do away with this aforementioned "discrimination."

The result of this exceptional sense of justice is not only a handsome wristwatch, but also a technical beauty with which fans of classical watchmaking are sure to fall in love.

As every human face is composed of eyes, a mouth, and a nose, and with these same ingredients both an ugly and a beautiful face can be created, so it is with watch movements. Every manually wound movement comprises plates, a spring barrel, a gear train, an escapement, and a balance. These parts can be put together with practicality in mind, or they can be supplemented with aesthetic additions—as is the case with the Cabaret's movement.

The clean, rectangular shape of Caliber L931.3 is without compromise; the lever for the large date correction cheekily juts out of it. Like a castle built to withstand a siege, the three-quarter plate and the escapement bridge protect the sensitive balance on three sides. The Glashütte ribbing diagonally decorating the plate of German silver creates a visual connection between it and the escape wheel bridge, located somewhat farther away. The large, deep blue screws holding the movement together emphasize its rectangular shape.

The somewhat larger engravings flushed out with gold are so cleverly spread over the three-quarter plate that the eye, when it can finally tear itself away from the pulsating balance, polished swan-neck fine adjustment, and screw-mounted gold chatons, very much enjoys what it sees when looking through the sapphire crystal case back of the Cabaret.

One of the advantages of the rectangular Cabaret movement is that even the layman (please excuse the use of this term; we are aware that only a connoisseur would buy this watch) or non-watchmaker can follow the flow of energy and the mechanical processes within the watch's movement well.

When looking at Caliber L 931.3 so that the engravings can be read horizontally, a large hole drilled into the plate can be seen to the right. Visible there is the click, looking for all the world like the beak of a bird of prey, three teeth of the ratchet wheel, and way to the right of the hole, the end of the click spring. Both click and spring are beveled and polished according to vintage

Cabaret Moon Phase

Attention to detail: the color of the thread used for sewing the leather strap matches the red gold of the case exactly.

The Cabaret introduced in 2004: its moon phase display builds a visual counterbalance to the outsize date.

Cabaret Moon Phase

A glance under the dial reveals the close quarters of delicate Caliber L931.5 with moon phase display.

CABARET Moon Phase

Manufactured since	2004
Movement	mechanical with manual winding, Caliber L931.5, 25.6 x 17.6 mm, height 5.05 mm, 31 jewels, 3 in screw-mounted gold chatons, glucydur screw balance, Nivarox balance spring, swan-neck fine adjustment, beat regulation via micrometer screw; stop-seconds, 21,600 vph, power reserve 42 h; Glashütte three-quarter plate, bridges with Glashütte ribbing
Functions	hours, minutes, subsidiary seconds; outsize date (patented); moon phase display
Case	dimensions 36.3 x 25.5 mm, height 9.1 mm, sapphire crystal, sapphire crystal case back secured with six screws, corrector button for date display on edge of case

Lange tradition. The interplay of these three parts ensures that the mainspring can be tensioned when the watch is wound.

The ratchet wheel in Lange watches is traditionally placed on the square hole of the spring's core (barrel arbor). This component revolves within a chaton secured by three screws to the far right. Next to it, slightly diagonal and also fastened with three screws, is the gold chaton containing the minute wheel's jewel. Next, and slightly diagonal, to this is the third wheel's jewel, and finally to the left of that and in a chaton held by two blued screws, exactly in the middle of the flat side of the movement, is the fourth wheel, carrying the small second hand on its long pivot at the other end of its stem.

Across from this, positioned like an island underneath the balance cock, the escape wheel revolves—a component still belonging to the gear train, although it is part of the escapement as well. With its help, the rotating motion of the gear train transposes its energy to the balance. To the right of the escape wheel bridge, under the balance ring, the pallets and lever are visible.

The elaborately engraved balance cock is a very complex component with which several important adjustments can be performed. Toward the bottom of the movement, jutting out above the balance cock, the satin-finished balance spring stud carrier that houses the triangular balance spring stud is visible. It can be adjusted to create just the right position for the balance in relation to the pallets (the so-called beat), after which it is then locked into place with a screw located on the balance cock.

The balance is encircled by a steel ring carrying the index on one side and the balance spring buckle across from it. The index is positioned on the balance cock and surrounded by the swan-neck fine adjustment spring. It is pushed against a little micrometer screw by the spring that gives this regulating organ its name, which juts out of the swan-neck spring on the side of the minute wheel in the direction of the balance cock. Because the little screw cannot be adjusted with a screwdriver due to its positioning, its head is drilled on the side to take on a fine regulating tip.

When the micrometer screw turns, the index must follow it due to the pressure of the swan-neck spring, also

Cabaret

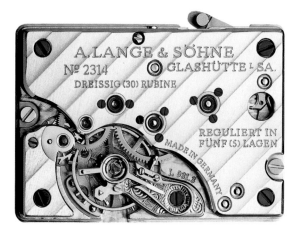

CABARET

Manufactured since	1997
Movement	mechanical manually wound movement, shaped Lange Caliber L 931.3, 25.6 x 17.6 mm, height 4.95 mm, 30 jewels, 3 in screw-mounted gold chatons, glucydur screw balance, Nivarox balance spring with swan-neck fine adjustment, patented beat adjustment via micrometer screw, stop-seconds, 21,600 vph, power reserve 42 h; Glashütte three-quarter plate, bridges decorated with Glashütte ribbing
Functions	hours, minutes, subsidiary seconds, outsize date (patented)
Case	36.3 x 25.5 mm, height 9.1 mm, sapphire crystal, sapphire crystal case back secured by six screws; corrector button for date display on edge of case

moving the regulator index, and once again changing the effective length of the Nivarox balance spring and influencing the watch's rate.

The results of this process are displayed on the large subsidiary seconds dial located on the flip side of the Cabaret, the silver-white of which against the black color of the dial seeming almost provocative. So that the watch can be precisely set, the movement is stopped when the gold crown is pulled to the hand-setting position. Above the crown, recessed on the right-hand side of the case, is located the button that activates the outsize date corrector. Its function is extensively described in the chapter concerning the Arkade model.

We end our treatise on the Cabaret with one last comment: This rectangular watch is certainly not a square!

Even More Beautiful with the Moon

A mild summer night becomes even more beautiful with a big, fat, full moon in a clear sky. Surprisingly, a moon phase display can also change the visuals of a very cleanly designed men's wristwatch in a positive way without influencing the desired "sober manly" look.

To the contrary: the classic moon phase display located within the small subsidiary seconds dial actually creates a visual balance on the Cabaret. Furthermore, it adds a splash of color to the dial with its dark blue night sky disk, pleasantly contrasting with the sober numerals of the dominant date display. The Cabaret with moon phase display, presented in 2004, is incredibly similar to its older sibling in looks. Only a second recessed correction button on the right side of the case reveals that there is more to set on this watch than just the date. ■

The Cabaret model is also available in a jeweled version featuring a paved dial and a white gold case set with 224 brilliant-cut diamonds.

Langematik

Modesty is a Virtue

The Langematik, launched in 1997, is the perfect example of an almost forgotten credo in our society: Still waters run deep. The outwardly reserved and unobtrusive watch turns out to be a veritable treasure chest of traditional horology and technical innovation when more closely observed.

From the classical design of the Langematik and its solid silver dial's conservative layout it is almost impossible to guess what kind of clever technology is hiding inside the case—available in white, red, and yellow gold or platinum.

The word "SAX-O-MAT" printed in place of the marker at four o'clock has an almost cheeky effect. SAX-O-MAT happens to be the name of the first automatic wristwatch movement to be issued from the house of A. Lange & Söhne. Pride in the company and a bit of regional patriotism jump out of the timepiece's name in equal parts, understandably concerning this watch—one that not only has an unusual name to offer, but also unique technology.

The movement's name is a conscious reminder of the half-automatic automobile engine Saxomat, made in the 1970s and '80s and successfully sold by Germany's Fichtel & Sachs, a company also formerly belonging to the Mannesmann concern. The new name SAX-O-MAT also contains a play on words—you only have to imagine the middle O as a zero to get it. This zero is of special importance in characterizing the newly developed automatic movement—so important that even the Saxons, so famous for their conscientious use of the German language, used an English-language name as an exception for their new patent: zero reset.

This new device is explained in the Lange catalogue a lot less simply, but far more precisely: "Hand-setting mechanism with automatic return of the second hand to zero." This basically means that the little second hand jumps directly to zero when the crown is pulled out to set the Langematik—an incredibly sensible invention, making the precisely synchronized starting of the minute and second hands possible (please see page 87 for more information).

Alongside this remarkable technical innovation, the Langematik model naturally also offers every well-known horological comfort. The topic of automatic winding is not a new one for A. Lange & Söhne: In the year 1891

Langematik

The Langematik in red and yellow gold cases and three dial variations. To the left, the impressively embossed three-quarter rotor with a screw-mounted platinum oscillating weight.

Langematik

Lange Caliber L921.4 SAX-O-MAT: It has a diameter of 30.4 mm and a height of 5.55 mm. Three-quarter plate made of German silver with Glashütte ribbing; three-quarter rotor (it takes up three quarters of the movement diameter) crafted in gold and platinum; four micro balls in ball bearing; hand-engraved balance cock; swan-neck fine adjustment; stop-seconds with automatic resetting of second hand (zero reset). The movement is regulated in five positions and its glucydur screw balance works at a frequency of 21,600 vph.

Emil and Richard Lange, the sons of company founder Ferdinand Adolph Lange, unveiled a self-winding pocket watch bearing the melodious name Perpetuale—even though this timepiece didn't come very close to the ideal of the *perpetuum mobile*. A watch with automatic winding needs movement, and in the pocket of a suit coat it wasn't moved much at all. In the twenty-nine years until 1920 only thirty-eight men wanted to own the Perpetuale model, making it one of today's most sought-after collector's pieces.

A Langematik wearer can even afford to be a couch potato, for the unusually designed automatic winding of the SAX-O-MAT works with very little friction and is highly effective due in large part to its very heavy rotor. ∎

LANGEMATIK

Manufactured since	1997
Movement	mechanical automatic movement, Lange Caliber L921.4 SAX-O-MAT, diameter 30.4 mm, height 5.55 mm, 45 jewels, glucydur screw balance, Nivarox balance spring with swan-neck fine adjustment, patented beat adjustment via micrometer screw, bilaterally winding embossed three-quarter rotor in 21-karat gold and platinum with 4 ball bearings in the reverser gear, 21,600 vph, power reserve 46 h; bridges decorated with Glashütte ribbing
Functions	hours, minutes, subsidiary seconds; stop-seconds mechanism with automatic resetting of the second hand (zero reset function) when the crown is pulled
Case	diameter 37 mm, height 9.7 mm, sapphire crystal, sapphire crystal case back secured by six screws; corrector button for date display on edge of case

Langematik

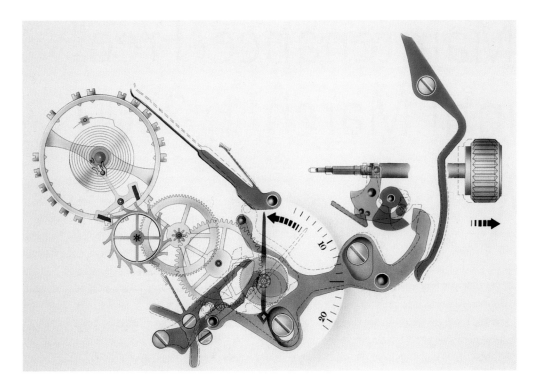

A schematic representation of the automatic resetting of the second hand (zero reset), denoting here a stopped watch (the dotted lines show the levers' positions when the movement is running). In the illustration, the winding stem (crown) is in the position to set hands. The angle lever (light blue) has shifted the transmission lever (grey) with the help of an additional small lever (dark blue). It pushes the reset lever (brown) against the flat side of the heart piece and simultaneously blocks the brake lever (purple) and the brake wheel (orange). At the same time, the thin balance blocking lever is pushed against the balance wheel by its own spring. The watch is stopped and the second hand jumps to zero.

Everything Stops at Zero

As with the tourbillon, where an enormously complicated mechanism had to be constructed in order to realize a comparatively simple idea, the zero reset function is a complex interplay of levers and springs that serve the almost banal goal of moving the second hand directly to zero when the Langematik's crown is pulled out.

Decisive for this function is a heart-shaped cam, simply called the "heart piece" by watchmakers. The heart piece, developed by Swiss watchmaker Adolphe Nicolet in 1862, is actually an important component for resetting chronograph hands. This resetting process was instituted nearly identically for the zero reset function as well. The function of the chronograph buttons was, however, taken over by the Langematik's crown.

And this is how zero reset works. The second hand of Caliber L 921.4 is perched on a separate stem driven by the fourth wheel located outside of the energy flow, and the heart piece is firmly attached to this stem. The stem of the fourth wheel is connected to the gear train by a clutch.

The angle lever (a control device located in the winding mechanism) is activated by the winding stem and places the movement in the winding and hand-setting positions. With the help of a powerful spring and an additional small lever, when the crown is used for winding it shifts a multi-legged transmission lever that is screwed onto the base plate of the movement.

This lever engages the reset lever, which in turn puts pressure on the heart piece and makes the second hand jump back to zero. The second hand returns to zero backward if it has not yet reached the 30-second mark, and jumps ahead to zero if it has passed the 30-second mark. At the same time, with a separate leg, the transmission lever also engages a very small stop lever that puts pressure on the balance and stops the watch. Now the hands can be set and the crown pushed back in. The balance and heart piece located on the stem of the fourth wheel are then immediately released.

In order to avoid the toothed wheels' necessary clearance being pushed back when the second hand rushes back to zero (from 29 to zero), a situation that would create a hesitation in the start function, the zero lever steers a blocking click with fine teeth, one that has a firm grip on the brake wheel located on the stem of the fourth wheel, before the actual zero reset process begins. Thus the fourth wheel and the second hand remain energized, and the Langematik model can immediately start when the crown is pushed in.

Langematik Perpetual

"Maintenance-Free" until March 1, 2100

A Lange watch containing a perpetual calendar was to be expected. And it came in the guise of the Langematik Perpetual in 2001.

With the advent of the Langematik Perpetual model, the manufacture only broke new ground by putting this complication into a wristwatch, for in the period between 1887 and 1931, pocket watches containing perpetual calendars were created under the auspices of Ferdinand Adolph Lange's sons and grandsons. In number these complicated watches only totaled five pieces, however, making them rarities of incalculable value today.

New, and thus different than all other watches made until now containing a perpetual calendar, is the combination of the Lange outsize date with the complicated mechanics of a calendar that contains—along with the date—the month, day, moon phase, and the four-year cycle marking one leap year to the next. Furthermore, this movement has been outfitted with a 24-hour display as well as the ingenious zero reset device, for this patented innovation was originally created for the automatic SAX-O-MAT Caliber L921.4. This movement serves as the base caliber for the Langematik Perpetual model and bears the new caliber name L922.1.

As if this weren't enough Saxon inventive spirit, Lange's designers and watchmakers added another special feature: the various displays of the perpetual calendar can be set independently of each other with separate recessed correction buttons located on the edge of the case. Additionally, all of the Langematik model's indicators can be changed simultaneously and in a synchronized fashion by activating the rectangular button located on the case at 10 o'clock, normally used for the quick adjustment of the outsize date on Lange's other watches. Moreover, the timepiece's hands can be adjusted anytime in either direction without the fear of possible damage to the calendar technology. It is normally not a good idea to adjust any functions while the calendar mechanism is engaged.

Clear Relationships

Because of the sheer amount of information that a watch containing a perpetual calendar must display, it would seem like a difficult task to place the various displays on a dial in a clean and legible manner. This problem was also cleverly solved by A. Lange & Söhne. The Saxons chose two of their own characteristic indications in the patented outsize date and the cleanly designed subsidiary seconds, giving them a dominant position on the dial relative to their importance. A remarkable fact is that the subsidiary seconds display, despite the moon phase indicator being integrated into it, has lost none of its clarity. The outsize date and the subsidiary seconds combined with the dial's powerful gold hands, large printed minute scale, and applied gold numerals allow the observer to gather basic information at a glance.

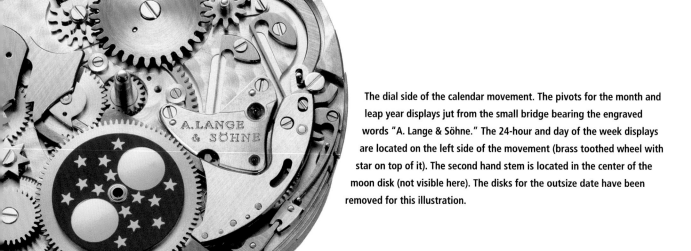

The dial side of the calendar movement. The pivots for the month and leap year displays jut from the small bridge bearing the engraved words "A. Lange & Söhne." The 24-hour and day of the week displays are located on the left side of the movement (brass toothed wheel with star on top of it). The second hand stem is located in the center of the moon disk (not visible here). The disks for the outsize date have been removed for this illustration.

The Langematik Perpetual with displays of date, day, month, leap year, and moon phase as well as an additional 24-hour display. The little picture shows how the second hand is set to zero by the so-called zero reset mechanism when the crown is pulled.

The other indications are placed just as neatly on the dial. But this is done so cleverly that the subdials at 3 and 9 o'clock seem to recede into the background.

At second glance one can immediately glean information about the current day of the week, the month, the cycles of day and night via the 24-hour display, or the answer to a question about when the next leap year will take place.

Even if the timepiece's owner wears it or winds it by hand regularly, all of the dial's displays will only remain correct until the end of February 2100. That's okay. For all of you owners out there: just change the date on your Langematik Perpetual to March 1 shortly after midnight on February 28, 2100 using the single correction button. And sleep well!

Celebrating Ten Years

Back in the day, parents and grandparents preferred to give their children and grandchildren a "good watch" for first communion or confirmation. Since these religious festivities aren't really the right thing for the gift of a watch from a true manufacture, however, A. Lange &

Langematik Perpetual

Illustration of the calendar mechanism with arrows representing the movement of each component's function. Especially interesting here are the square-toothed disk (purple) for the display of the months' lengths over the course of four years and the snail-shaped cam and hook-shaped lever resting on it (to the right underneath the moon disk, both colored blue) ensuring that the date window displays the first of a new month.

Eternity is Not Forever

Early clocks also satisfied a general interest in information regarding additional displays above and beyond the actual time—such as the date and moon phases. Even some antique water clocks already showed certain astronomical events. In more modern times enormous "time display machines" were created with astronomical dials that could be found above all in churches. The Strasbourg cathedral, the cathedral of Exeter, and the newly restored clock of San Marco in Venice are just some examples of this.

Contemporary watches featuring calendar displays are categorized as follows: A simple day-date display that needs to be adjusted at the end of a month containing less than 31 days; another containing a so-called complete calendar that not only continually shows the correct date for a long period of time, but also the month and often the moon phase; and yet another containing a so-called perpetual calendar that also keeps track of leap years as well as the time, day, date, month, year (usually displaying the leap year cycle by using the numerals one through four), and the moon phase until March 1, 2100.

The previously mentioned year 2100, which not one of the proud owners of one of these complicated watches will experience firsthand, is a so-called secular year. According to the Gregorian calendar that we follow, we have a leap year every four years, but only in the first year of the new century if it is divisible by 400. Because a secular year is therefore not a leap year, the 29th of February will not exist in that year.
There is no such thing as a wristwatch that can take secular years into consideration. Not even one made by A. Lange & Söhne.

LANGEMATIK PERPETUAL

Manufactured since	2001
Movement	mechanical automatic movement, Caliber L922.1 SAX-O-MAT, diameter 30.4 mm, height 5.7 mm, 43 jewels; glucydur screw balance, Nivarox balance spring with swan-neck fine adjustment, patented beat adjustment via micrometer screw, bilaterally winding embossed three-quarter rotor in 21-karat gold and platinum with 4 ball bearings in the reverser gear, 21,600 vph, power reserve 46 h; bridges decorated with Glashütte ribbing
Functions	hours, minutes, subsidiary seconds; perpetual calendar with outsize date, day, month, moon phase, and leap year indication; 24-hour display with day/night indication; stop-seconds mechanism with automatic resetting of the second hand (zero reset function) when the crown is pulled
Case	diameter 38.5 mm, height 10.2 mm, sapphire crystal, sapphire crystal case back secured by six screws; recessed corrector on edge of case

Anniversary Langematik

Söhne chose another reason for celebration to do a select number of watch connoisseurs a great big favor. The prices for this company's products aren't exactly what one would spend for a communion watch anyway—unless one had a first communion every year, and the child was meanwhile thirty!

The tenth anniversary of Walter Lange's registration of Lange Uhren GmbH in 1990 was, understandably, a wonderful occasion for making a special limited edition of the Langematik available only in platinum. While the valuable, steel-grey case material fits in well with the concept of understatement, something the manufacture and the lion's share of its clientele value greatly, the red Roman twelve on the enameled dial seems almost daring.

Enamel was also the classic material for fine pocket watch dials and lent the Anniversary Langematik a special elegance with its unmistakable splash of color. As did Caliber L921.2 with its rotor made of platinum and gold and the other characteristics of the individual movement embellishment that have come to be expected of a watch by A. Lange & Söhne.

The 500 of these beautiful, classically elegant watches, serially numbered, are of course long since sold out. But in just a few years Lange Uhren GmbH will be celebrating its twentieth anniversary—certainly with another anniversary watch. ∎

The Anniversary Langematik with platinum case and automatic manufacture movement L921.7 SAX-O-MAT has been long sold out. This should surprise no one, for this watch—manufactured in Lange's inimitable way down the last screw—differs in subtle ways from the other company products. Perhaps this is due to the hand-enameled dial that gives this classic men's watch a pinch of high spirits with its red XII. Perhaps it is also the missing outsize date, which isn't really missed on this anniversary watch.

Datograph

The Outsize Date and the "Time Writer"

A worldwide innovation presented at the Basel Fair of April 1999, the Datograph model sets new heights in both appearance and technology.

In 1868 A. Lange & Söhne was already making pocket watches with chronographs. This collector's piece is from 1903 and bears the movement number 45.697.

Not only does the technology hidden under the dial have no equal, the face of this "time writer" (a literal translation of the Greek word chronograph) is also unusually attractive.

The metal plate of solid silver making up the Datograph's delightful "visage" comprises both the actual dial and the encircling ring bearing the tachymeter scale and the words "A. Lange & Söhne." The dial designers were especially clever with the two asymmetrically arranged subdials for the display of subsidiary seconds and the 30-minute counter, setting a visual antithesis to the dominant outsize date ("dato"-graph) located in the upper half of the dial. In this way, the date display is integrated into the very technical appearance of the dial in a masterly fashion.

The modern-styled case featuring powerful lugs and a large crown is offset with classic elements like the lance-shaped hands and the Roman numerals 2, 6, and 10 on the dial. At the same time, the simple line markers denoting 1, 5, 7, and 11 avoid any old-fashioned feeling.

The fact that the designers ignored the unwritten rule that dials with hour markers formed by a mixture of numerals and indexes should put the numerals at the quarter hours, on the one hand, is necessary because of the position of the subdials and, on the other hand, serves to enhance the unusual attractiveness of this timepiece.

Time Writer without Words

The word "chronograph" is made up of the Greek roots *chronos* (time) and *graphô* (I write). Of course, it is not possible to write down the time, and so the first "time writer" from 1822 only served to document the time graphically by marking a time interval on a dial made of paper with the help of a hand bearing a type of ink pen.

In the year 1862 Adolphe Nicolet from Switzerland's Vallée de Joux developed the first chronograph containing a resettable hand and using the same heart piece to set the hand to zero that the Datograph model uses. In 1868 Ferdinand Adolph Lange, founder of the Lange & Söhne manufacture, produced the first pocket watch chronograph using a so-called column wheel type of construction. This term is derived from the French word *colonne* for column.

The Datograph was only available in platinum at first; in 2005 an elegant red gold version was introduced.

By the year 1868 no one was speaking of "time writing" anymore, and chronographs had become what they are today: wristwatches with an integrated stopwatch powered by the watch's movement.

Triangular Columns

The Datograph is also outfitted with a movement that uses column wheel technology, still considered the luxury class in chronograph mechanisms. Using computer terminology, the column wheel is of course more similar to hardware in the literal sense (it is made of hardened steel), but functions like a software program within the movement. All of the control operations in the watch's stop mechanism are dependent upon the column wheel. It even functions on a simple on/off principle, the same basis used by all of the complicated operations in a computer.

By activating the "on" button, located at 2 o'clock on the case, the starting lever receives the order to advance one tooth further. To carry out this order, a small hook on the lever engages the column wheel's teeth, moving it one tooth further, after which a powerful brake spring immediately secures the new position.

With the aid of its triangular "columns," the column wheel alternately orders the so-called clutch to engage and disengage. The clutch, with which the connection to the gear train is made, is a type of moveable wheel bridge to take on the transmission wheel. If the clutch receives the order to engage, it slides its feeler spindle between two of the seven triangular columns making up the column wheel. This little motion is enough to make a connection between the wheels of the movement and the chronograph.

The first part of the process is repeated when the start button is pushed again, but the clutch's feeler spindle is raised this time by one of the triangular columns to break the connection between the transmission wheel on the clutch and the chronograph's wheel to count seconds. The chronograph hands remain in the same position.

Double the Fun with Two Fourth Wheels

The chronograph mechanism is usually driven by the fourth wheel. This is a natural configuration, for a 1:1 translation is automatically achieved by the chronograph wheel located in the center of the movement and rotating once a minute.

There are various ways of transferring the energy of the fourth wheel rotating between the plates of the original movement and bearing the hand of the subsidiary seconds display to the chronograph device located on the rear gear train bridge, outside of the movement. A popular method, for example, is to outfit the stem of the fourth wheel rotating within the movement with two long pivots. One is necessary anyway on the dial side for the small second hand. A double fourth wheel is placed on these long pivots, and is now called a carrier wheel. Taking the carrier wheel back off is always a bit of a risk for watchmakers, for the hard pivot upon which it firmly sits can easily break off. A. Lange & Söhne has now helped the watchmaker in this situation. Caliber L 951.1 is also outfitted with this double fourth wheel solution, but its upper (for watchmakers the back of the movement is always up) fourth wheel is positioned differently. This movement is outfitted with its own fourth wheel bridge, and for this reason one rotates between the plates, and a second rotates outside of the lower part of the bridge where no possible damage can be done by a watchmaker wanting to remove it from the pivot. Installing the escape wheel bridge is somewhat more complicated than on a normal construction, however, as it has to be maneuvered between the wheels positioned on the stem of the fourth wheel.

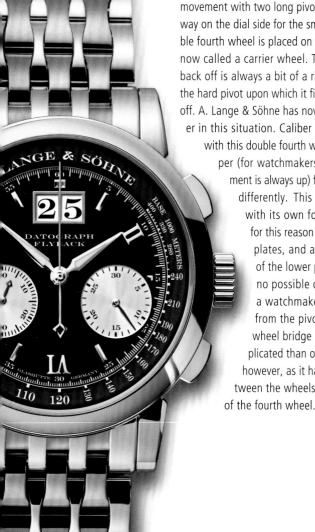

Resetting on the Fly

The realization of the flyback function demands of the movement designer more imagination than the nimble jump of the chronograph hand on the dial might reveal.

The less usual French term for this type of watch is the *retour en vol*, while German-speakers may or may not prefer "permanent reset." Although it may sound like it, this function actually has very little to do with flying. The terms are referring to the fact that the chronograph can be reset to zero while still running, without first being stopped. Or, to put it a different way, that the sweep chronograph hand (chronograph second counter) and the minute counter hand can be interrupted while making their rounds, "flying" back to their reset points lickety-split. For form's sake, it should also be mentioned that the hands, depending on the corresponding wheels' positions, can also "fly" in a clockwise direction (flying "forward" instead of "back").

In order to achieve this, the chronograph must be designed so that the operating lever, which is activated by a button at 4 o'clock that is normally blocked when times are being measured, can also be moved when the chronograph is running. When this happens, the two-legged chronograph heart is flung toward the center of the movement. The slanted ends of the chronograph heart thus meet with two heart-shaped cams situated on the stems of the chronograph center wheel and the minute counter's wheel. The heart cams then slide abruptly along the chronograph heart, turn their flat sides against its slanted surfaces, and remain there. On the front of the watch, this is perceived as a jump of the corresponding hand to its reset position.

Normally, the wheels would remain in this position until the next time that the start button at 2 o'clock is activated. This would, however, preclude the reset lever, controlled by the column wheel, from being engaged, but this is something that should be avoided at all costs. It is better that the chronograph heart jumps right back into the position that it would normally have when the stop mechanism is running (start position). On a flyback, this happens as soon as the button at 4 o'clock is let go. The freed-up wheels once again rotate, and the hands for the second and minute totalizers go right back to work.

In order to make this changed process possible, watch technicians must design the chronograph mechanics so that on one side the start-stop-reset process is possible in that order and, on the other side, the "stop" step can be alternatively passed over, making the "start" function superfluous.

Datograph

The illustration on top makes the effect of the permanent reset (flyback) clear. When the large arced lever to the far right (the so-called reset lever) is activated, a second lever (called the heart lever because it pushes the heart piece), whose legs in profile remind one of a ski jump, moves into position on the dotted line. This breaks the connection between the rotating transmission wheel located in the clutch lever (underneath the triple screw-mounted chaton) and the wheel for the second counter. Almost simultaneously both of these counter wheels are reset by the previously mentioned double-legged heart lever and are able to immediately move again. The wheels can start again in the blink of an eye and begin measuring anew.

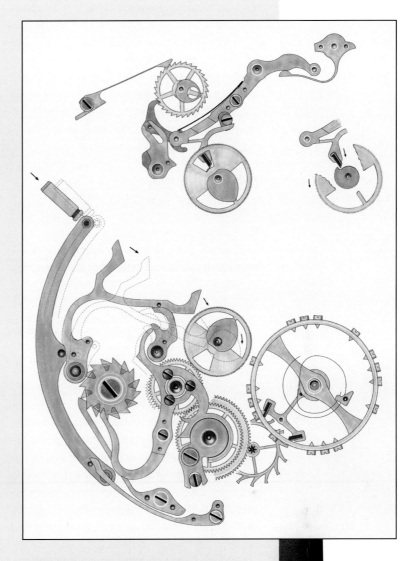

The connection between the permanently rotating wheels of the movement and the temporary motion of the chronograph mechanism is guaranteed by the traversable clutch lever. Necessary to this is the chronograph clutch wheel that creates the connection between the movement on one side and the second and minute counter wheels on the other via a transmission wheel located on the fourth wheel's stem that is located on the clutch lever supplying the energy from the movement to the chronograph.

Controlled by the column wheel, the clutch swivels back and forth around the axis of the fourth wheel, though only a few degrees, and is located on a screw or another immovable part. The teeth of the clutch wheel and chronograph wheel thus engage each other or are separated.

The clutch usually swings back and separates the toothed wheels joining the chronograph and the movement as soon as the stop function is chosen. Only then can the chronograph heart activate without danger to the gear train.

On the flyback mechanism, the clutch only has fractions of a second to swing back before the control surfaces of the chronograph heart touch the heart cams.

And the chronograph user doesn't notice a thing. He or she only sees two hands that "fly" back to their reset points.

The illustration on the bottom shows the mechanism of the precisely jumping minute counter. To the left, you can see how the jewel slides from the snail cam's step. Using the energy created by this, the toothed wheel is suddenly rotated one tooth further by the claw of the minute counter lever. The wheel for the second counter now rotates counterclockwise (because we are looking at the back of the movement). The minute counter lever (a two-legged lever, jeweled to avoid friction) is permanently pushed against the snail cam underneath the wheel by the spring. At the same time, it is slowly "lifted" and the claw can latch onto the next tooth of the minute counter's "circular saw." This process continues to repeat.

Datograph

DATOGRAPH

Manufactured since	1999
Movement	mechanical manually wound movement, Lange Caliber L951.1, diameter 30.6 mm, height 7.5 mm, 40 jewels, 4 in screw-mounted gold chatons, 18,000 vph, glucydur screw balance, stop-seconds, patented beat regulation, Nivarox balance spring, column-wheel control of chronograph functions; Glashütte three-quarter plate, bridges decorated with Glashütte ribbing; movement decorated and engraved by hand
Functions	hours, minutes, subsidiary seconds, outsize date (patented); chronograph with flyback function and precisely jumping minute counter
Case	diameter 39 mm, height 12.8 mm, sapphire crystal, sapphire crystal case back secured by six screws; corrector button for date display on edge of case

Sporty Minute Counter

On most chronographs, the wheel for the minute counter positioned underneath the dial's 30-minute totalizer is usually activated by a carrier located on the wheel for the second counter rotating once. The fact that this process, taking place once a minute, lasts for several tenths of a second was apparently unacceptable for A. Lange & Söhne. This is why head designer Helmut Geyer and his team created a complicated mechanism, one that was already known to watchmaking in principle, but which no one had ever achieved miniaturizing, that causes the small minute hand to jump abruptly from one marker to the next every 60 seconds.

The main component of this complicated technology is a two-legged lever—the function lever controlling the minute counter. Similar to the start lever when it is activated, it bears a movable claw on one end, using it to grip between the teeth of the minute counter's wheel.

A stepped fusee is located on the stem of the wheel for the second counter (rotating completely once a minute). The two-legged minute counter control lever is pushed gently against this stepped fusee by a spring, led by its feeler made of synthetic ruby. Because of the eccentric shape of the stepped disk, the control lever is pushed further and further away during the one-minute rotation of the wheel for the second counter. During this process the control lever spring is continuously tensioning, and the lever's claw slips behind the next tooth of the minute counter's wheel. After one minute passes, the ruby feeler falls from the fusee's step, the lever is pushed in the direction of the second counter's wheel, and the claw turns it.

The minute counter's wheel and its control lever move in jeweled beds, of which the upper ones—visible from the outside—are located in triple screw-mounted gold chatons. More gold chatons serve as beds for jewels of the second counter's wheel and transmission wheel located on the clutch lever.

Flyback is the name of another, seldom-used additional function with which Caliber L951.1 is outfitted (please see pages 94-95 for more information).

One Can Do Without Anything But Luxury

It's fairly obvious that this statement could be referring to any product by A. Lange & Söhne, and the Datograph is no exception. Let it be said however that this watch's balance cock is also hand-engraved, the escape wheel turns on jewels with additional endstones that are bedded in finely polished, double-screwed endpieces, and that the balance, outfitted with a spring the quality of Nivarox 1, can be regulated with the aid of a swan-neck fine adjustment mechanism.

Even the start lever of this chronograph moves on jewels and may, along with the other beautiful elements and many functions this movement has to offer, be admired through the sapphire crystal case back of the case. ■

Double the Fun

Lange Double Split

The Lange Double Split by A. Lange & Söhne is the first chronograph ever to offer a practical rattrapante function for both the second and the minute counters.

Until the introduction of the Lange Double Split, the possibility of individually stopping and restarting the two concentrically running hands of a rattrapante chronograph was restricted. The length of time periods measurable by stopping the two hands was limited to intervals of 59 seconds. Of course, it was possible to just leave the second hand where it was for more than a minute while the other hand continued to run. But then it was no longer possible to tell whether a time of 37 seconds or four minutes and 37 seconds had elapsed.

With the Double Split, A. Lange & Söhne has now included the minute as the next largest unit of measurement in interval timing results. Several years of developmental work helped Lange's movement designers realize a chronograph whose 30-minute totalizer is also outfitted with two hands that function independently of each other—though always in synchronization with the second hands—that can be started, stopped, and once again synchronized.

The Lange Double Split can measure the usual 30-minute intervals, but along with it also the results of intermediate timing up to 29 minutes and 59 seconds.

"Split" Seconds

A catchy, multilingual name was conceived for the Datograph when it was introduced seven years ago, a timepiece whose movement partially provided the base for Caliber L001.1. This name hints at both the patented Lange outsize date and the stopwatch function it is outfitted with. The Lange Double Split is the first of A. Lange & Söhne's wristwatches to be christened with a purely English name.

For a good reason, as it turns out. The words "split seconds" are used in the global language English—a language that most of Lange's customers speak—to describe this chronograph's specialty. It is called a *rattrapante* by French-speaking watch fans, and—somewhat less elegantly—a *Schleppzeiger* or double chronograph by the Germans.

Doubled on this type of watch is always the set of chronograph second hands and their corresponding wheels, whereby the stem of the first fourth wheel (chronograph wheel) rotates in the drilled stem of the center wheel, and that of the second counter wheel (rattrapante wheel) in the stem of the first one.

The Lange Double Split is the first flyback chronograph worldwide with a double rattrapante, controlled by classic column wheels.

Lange Double Split

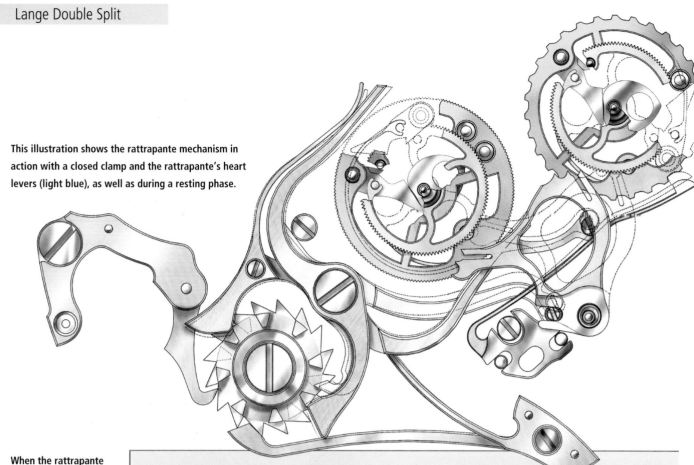

This illustration shows the rattrapante mechanism in action with a closed clamp and the rattrapante's heart levers (light blue), as well as during a resting phase.

When the rattrapante function is activated, the rattrapante heart lever usually scrapes across the heart pieces, which can lead to energy loss in the movement. The Lange Double Split is thus outfitted with a patented isolator that pushes back the heart lever when the rattrapante is activated.

Trailing Hand

Rattrapante, split-seconds, or just plain double chronograph—these are all names for the same thing. "Double" on these watches is the chronograph second hand, making rotations in a sweep manner from the center of the dial. After the stop mechanism has been activated, the "double" hands run congruently to each other and look for all the world just like one hand when glanced at fleetingly.

On these watches, the wheels belonging to the chronograph second hands are also always "double," though the stem of the first fourth wheel rotates in the drilled stem of the center wheel, and that of the second counter wheel in the stem of the first one. Both wheels run in a synchronized fashion when the normal chronograph is being used, and the hands driven by them are in the same position.

The technical specialty of this type of watch is that one of these two hands, independent of the other, can be stopped to measure an interval time by activating an additional button on the left side of the case or in the crown.

When this button is pushed again, the hand that had previously been stopped hurries to catch up with its twin brother, who has continued to run, for which reason it is called a *rattrapante* (French rattraper = to capture, to catch up with, even to make up for lost time), and now runs once again congruently to the other hand.

As long as both hands run together, the rattrapante's wheel is given a lift by a heart-shaped cam that is located on the chronograph wheel (second counter). A small, spring-loaded lever pushes again this heart cam, which can engage it on the flat side of the heart or by using a tiny roll between the heart cam's legs. Then both second hands are located on top of each other and start going together when the start button is pushed because the pressure of the little spring on one of the wheels is great enough to take the other one (on the heart cam) along.

When the button on the case for the rattrapante is activated, tongs controlled by the second of the chronograph's column wheels surround the toothless rattrapante wheel and hold it tight. The chronograph wheel continues to run, whereby the above-mentioned little spring-loaded lever, gently pushed against the heart-shaped cam, slides along it.

When the rattrapante button is pushed again, the spring-loaded lever pushes the roll to the "lowest" point of the heart cam, between its two legs or on the flat side of the heart cam, and the rattrapante wheel is put back in its original position. Then both of the second hands are once again congruent.

Lange Double Split

LANGE DOUBLE SPLIT

Manufactured since	2004
Movement	mechanical with manual winding, Caliber L001.1, diameter 30.6 mm, height 9.45 mm, 40 jewels, four in screw-mounted gold chatons, shockproofed glucydur screw balance with eccentric regulating screws, in-house balance spring, 21,600 vph; chronograph functions controlled by two column wheels; isolator mechanism; Glashütte three-quarter plate; bridges with Glashütte ribbing; movement decorated and engraved by hand
Functions	hours, minutes, subsidiary seconds; flyback chronograph with rattrapante for second and minute counters (precisely jumping); power reserve display
Case	platinum, diameter 43.2 mm, height 15.3 mm, sapphire crystal, sapphire crystal case back secured by six screws

Both wheels run in a synchronized fashion when the normal chronograph is being used, and the hands driven by them are in the same position (please see page 98 for more information).

The use of a split-seconds chronograph is easily explained: Two athletes, for example, are supposed to run a certain distance. In order to determine the varying amounts of time that both of them need to cover the distance, one would normally need two chronographs—or a chronograph outfitted with a rattrapante function. The use of one timepiece has the added advantage that there can be no distortion of the race's timed outcome due to human delay in pushing the two start buttons. When the race's starting signal sounds, the start button at 2:00 on the watch's case is pushed, and both athletes and both hands are off. When the first runner crosses the finish line, the button on the left side of the case is pushed and the first hand stops running. When the second runner reaches the finish line, the chronograph is stopped as usual, the second hand stops, and the times of the two competitors can be easily read. The only disadvantage would normally be the above-mentioned restriction of sixty seconds, but this has since become history with the introduction of the Lange Double Split.

Resetting on the Run

When a normal split-seconds function is activated, the chronograph heart drags across the controlling heart cam, which leads to loss of power in the movement and, as a result, a smaller vibration of the balance. Normally.

That is not so with this original Glashütte movement design. As is so often the case at Lange, a complex solution was devised for a relatively small problem, something that seems to be natural for this brand.

Both of the wheels of the second and minute counters are outfitted with a mechanism that separates the chronograph heart from the heart cam for a moment after the rattrapante has been stopped. This so-called isolator thus allows one to use the rattrapante function as much as desired without causing any additional energy

Lange Double Split

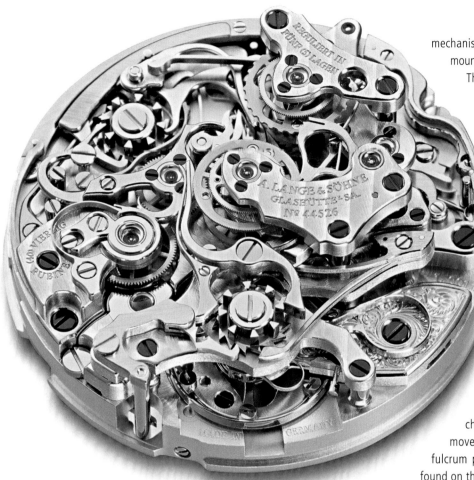

Caliber L001.1 is based on the Datograph's movement. It comprises 465 individual parts and has forty jewels of which four are screw-mounted in gold chatons. The balance beating at 21,600 vph is outfitted with an in-house balance spring.

mechanism for the second and minute counters is mounted onto the original chronograph module. Thus, a movement of imposing height and surprisingly airy design was created, opening an unusual new perspective. When looking through a loupe, an indispensable utensil that every watch lover certainly has on hand, one feels as if one has been placed in the middle of a giant machine in this mechanical microcosm. It's an impressive view even for connoisseurs!

The writing on this movement has once again been executed only in German. The movement is "regulated in five positions" and outfitted with "forty jewels" as announced by delicate engravings flushed out in gold on the bridge of the minute counter and the cock of the driving wheel, which unites the chronograph mechanism with the base movement. Concentrically positioned to it, the fulcrum point of the chronograph clutch lever is found on the cock, which touches the chronograph's column wheel on the opposite side with a feeler. On the other side, it allows the chronograph mechanism to participate in the flow of energy via the driving wheel that is positioned upon it in a normal jewel bearing on the bottom and a screw-mounted gold chaton on the top when its feeler is engaged between the column wheel's triangular columns. The clutch lever is engaged by a circular function spring that sits atop the cock along with the rotating element of the almost circular clutch lever, bringing it to the place required by the column wheel with its tension. The spring and the rocking rotational element thus build a visual as well as a functional unit, once again confirming the talent of A. Lange & Söhne's movement designers and watchmakers in creating a remarkably detailed synthesis of technical issues and aesthetics.

loss to the gear train, escapement, and balance. Alongside this load of innovative watch technology, it is almost easy to overlook another useful function of the Lange Double Split, which makes it especially attractive for users in a hurry: the flyback option that is especially loved by pilots, which allows one to reset the chronograph hands without first stopping them. On the Lange Double Split one can even use this function when the split-seconds hands are stopped; they just remain in position.

A Synthesis of Technology and Aesthetics

The sapphire crystal on the case back of the heavy platinum case—on a leather strap this watch weighs a hefty 220 grams—allows a look into the unique movement design of Caliber L001.1 (00 stands for the year that the company began working on it, which means 2000), actually inviting the observer to do so with its wonderfully anti-reflective surface.

The observer is greeted with a seemingly confusing scene at first, with roles taken over by innumerable levers, springs, and wheels. Almost all of the rattrapante

Not Only Special Mechanics

This successful mixture does not only restrict itself to the construction of the 9.45 mm high watch movement. The bezel located on the case back that has been especially decorated to hide the watch's stately height of 15.3 mm becomes a visual part of the case and is a successful mix-

1815 Chronograph

ture of elegance and technical grace. Its matte surface, from which the polished letters jump out like relief, perfectly matches the polished steel parts of the movement that it frames.

The front view of the chronograph is dominated by the deep black dial with extra-large subsidiary dials for the display of seconds and the minute counter, underscoring the character of this instrument watch. The applied white gold Roman numerals and hour markers as well as the decorated ends of the chronograph hands spice up the absolute sobriety propagated by the plain platinum case with a pinch of playfulness and elegance.

The outsize date display, which has meanwhile advanced to a hallmark for the A. Lange & Söhne brand, was purposely left off the Lange Double Split's dial, mainly for technical reasons. The room that it would have taken up on the timepiece's bipartite solid silver dial has been co-opted by the small indication informing the wearer of the watch's remaining power reserve. The completely typical A. Lange & Söhne arrangement of the dial, in which three displays form the points of a right triangle, is also evident on the Lange Double Split.

1815 Chronograph: The Traditional Chronograph

In 1868 A. Lange & Söhne was already making technically demanding pocket watches outfitted with stop functions in masterpieces that set the height of the measuring stick. Despite this, loyal to the company's leitmotiv of "progress and tradition," A. Lange & Söhne gave itself five years to prepare for the introduction of a "normal" chronograph after the sensational introduction of the Datograph in 1999—although the word "normal" is in no way a fitting attribute for the 1815 chronograph. It has the famous Datograph chronograph movement with classic, complex column wheel technology and a precisely jumping minute counter that additionally disposes of the rather rare flyback function. This function allows the running chronograph to reset and immediately begin again by the pushing of a button located at 4 o'clock.

The 1815 Chronograph, whose Caliber L951.0 is not outfitted with Lange's patented outsize date, and according to the company "reminds one of the stylistics of vintage pocket watch tradition like no other," is available in white or red gold and has on the edge of its solid silver dial a pulsometer scale with which it underscores its individuality along with subsidiary dials that are unusually far below the dial's horizontal. ■

1815 CHRONOGRAPH

Manufactured since	2004
Movement	mechanical manually wound movement, Lange Caliber L951.0, diameter 30.6 mm, height 6.1 mm, 34 jewels, 4 in screw-mounted gold chatons, 18,000 vph, glucydur screw balance, stop-seconds, patented beat regulation, Nivarox balance spring, column-wheel control of chronograph functions; Glashütte three-quarter plate, bridges decorated with Glashütte ribbing; movement decorated and engraved by hand
Functions	hours, minutes, subsidiary seconds, chronograph with flyback function and precisely jumping minute counter
Case	diameter 39.5 mm, height 10.8 mm, sapphire crystal, sapphire crystal case back secured by six screws

Datograph Perpetual

One-Fifth of a Second until Eternity

The Datograph Perpetual unites the practical use of a flyback chronograph with the advantages of a perpetual calendar displaying the date, day of the week, month, year, and moon phase.

Alongside the credo "nothing is so good that it can't be improved," a motto that A. Lange & Söhne truly breathes to life, another credo seems to have taken hold of the engineers and watchmakers at the manufacture: "nothing is complicated enough." Whereby one must always keep in mind that the term "complication"—in the sense of a display complication—is extremely positive when used in conjunction with valuable watches of the top luxury segment. Other things that are complicated on timepieces like the Datograph Perpetual are of course the movement design and the production of the movement, in this case the complexity of the new Caliber L952.1.

Extensive and correspondingly difficult to achieve are the technical processes within such a movement. Not only is it hard work for a mechanical watch movement to guarantee the function of the many continuous display mechanisms working together, but it must do so while simultaneously delivering a satisfactory rate performance.

This is true for the functions of the perpetual calendar as well, whose displays are correctly preprogrammed until the year 2100. This might seem a paradox in a watch that is manually wound, but it would be technically possible for the displays to be correct even longer if it weren't for the Gregorian calendar setting restrictions on horology. According to our calendar, the year 2100 is a secular year; it should by all rights be a leap year, however the 29th of February will not take place that time around. The leap year usually happens according to the calendar created by Pope Gregory XIII in 1582 every four years, but in the first year of a century it only occurs when the number is divisible by 400.

The Datograph Perpetual's calendar control mechanism does take leap years into consideration, shown on a small subsidiary dial located between 4 and 5 o'clock, but on March 1, 2100 a small manual correction, quickly completed by pushing the button on the date corrector, is unavoidable.

This is, however, not a burden for the watch's wearer, for Lange's movement designers have developed a new corrector mechanism for the 223-component strong calendar: By using teeth of varying lengths and a specially shaped cover on the calendar's control wheel, only one single click is needed. This arrangement, registered for a patent, has a shorter path to the control mechanism when being manually adjusted and thus makes easy changing of the outsize date possible.

Complicated Combinations

Neither in the development of traditionally designed chronographs with exclusive and practical additional functions nor in regard to a perpetually correct calendar does A. Lange & Söhne need to prove its know-how: The company already realized a perpetual calendar in 2001 in the form of the Langematik Perpetual, while the Lange Double Split and Datograph models have brought forth

Datograph Perpetual

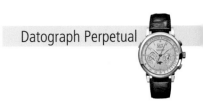

the most unusual chronographs from the Saxon manufacture in recent history.

The Datograph Perpetual is based on the two abovementioned models, and the most striking elements of those innovative timepieces are combined here. The displays of the calendar indications and the moon phase, which works so precisely that it will only deviate from the real moon by one day after 122 years, have been completed in the new chronograph with the addition of a day/night indication. Its wedge marker—in truth a rotating disk moved by a wheel as previously seen on the Lange 1 Time Zone—makes its rotations around a subsidiary dial on the upper edge of the scale reserved for the subsidiary seconds. In its center, the indication for the day of the week has also found room, underscored by being partially recessed within the rhodium-plated solid silver dial. The month display in the middle of the 30-minute totalizer subdial is similarly arranged. In this way, the unavoidable loss of legibility due to the combination of chronograph and perpetual calendar has been kept as slight as humanly possible.

This complicated new model has inherited valuable elements from the Datograph, such as the flyback mechanism and the precisely jumping minute counter, which are controlled by a column wheel (as if there would be another way at Lange). The new calendar chronograph allows interval measurements of up to thirty minutes that it can stop down to one fifth of a second.

Beautiful All the Way Around

Like all timepieces by A. Lange & Söhne, the Datograph Perpetual is also almost as beautiful from the back as from the front. This will hardly surprise the connoisseur, for the new model shares outstanding characteristics with the other Lange watch families, such as the flat movement components being made of German silver in its natural state, screw-mounted gold chatons, and finely polished steel parts.

There is, however, one thing that expresses the gaining autonomy of A. Lange & Söhne that may not be recognizable right off the bat through the sapphire crystal case back: The 18,000 vibrations per hour that the glucydur balance, finely adjusted by regulating screws, make are controlled by a balance spring produced in-house. Rate precision is also enhanced by a Lange development: The special position of the toothed wheels on the spring barrel limiting the winding and loosening of the mainspring so as to produce almost constant torque.

The dial side of Caliber L952.1, comprising 556 individual components, looks exceptionally clean and tidy—for a perpetual calendar, that is. That may be

DATOGRAPH PERPETUAL

Manufactured since	2006
Movement	mechanical with manual winding, Caliber L952.1, diameter 32 mm, height 8 mm, 45 jewels, four in screw-mounted gold chatons, shock-proofed glucydur screw balance with eccentric regulating screws, in-house balance spring, stop-seconds; 18,000 vph; chronograph functions controlled by column wheel; Glashütte three-quarter plate; bridges with Glashütte ribbing; movement decorated and engraved by hand
Functions	hours, minutes, subsidiary seconds; perpetual calendar (outsize date, day, month, moon phase, leap year); 24-hour display with day/night indicator; flyback chronograph with precisely jumping minute counter
Case	diameter 41 mm, height 13.5 mm, sapphire crystal, sapphire crystal case back secured by six screws; correction button for date display on edge of case

because this side of the new movement is naturally fully dominated by large-surfaced disks for displaying the patented outsize date.

The practical mechanical miniature work of art is housed in a platinum case with a diameter of 41 mm and a height of 13.5 mm. The hour and minute hands are inlaid with luminous substance. A typical product of A. Lange & Söhne: Practicality never hurts. ∎

Pour le Mérite

The Rocket and the Whirlwind

A. Lange & Söhne finds its way back to vintage design and watchmaking tradition, presenting a wristwatch of the superlative sort that masters both the slow motions of the spring barrel and the industrious whirlwind of the tourbillon: the Tourbograph Pour le Mérite.

However, this attractive set of words could only be put together because of the tourbillon that is the decisive characteristic of the Tourbograph Pour le Mérite. This complicated "mechanism for minimizing the influence of gravity in vertical positions of the watch movement," soberly dubbed a *Drehgang*, or *Karussell* in traditional Saxon watchmaker parlance, was called a whirlwind (tourbillon) by its inventor Abraham-Louis Breguet in his flowery mother tongue, French.

The second part of the model's name is easy to categorize: it comes from the word "chronograph." The two sweep second hands, one piled on top of the other, make it easy to see that this is indeed a chronograph of the best kind, including a split-seconds function (rattrapante).

Pour le Mérite, named after a Prussian order (French for "for the merit"), is a subtle homage to Ferdinand Adolph Lange from the company's master watchmakers. The idea for this amazing new watch introduced in 2005 apparently came from Günter Blümlein and former technical consultant Reinhard Meis in the 1990s according to sources in Glashütte.

The Tourbograph Pour le Mérite is thus dedicated to Blümlein who, in 1994, already introduced a Lange tourbillon by the honorable name of Pour le Mérite.

The Perpetual Goal—Rate Precision

Lange's movement designers and especially Annegret Fleischer, who was responsible for the project, once again proved their technical inventiveness and the competence necessary to realize this truly amazing watch, which was conceptually born in the workshops of Swiss think tank Renaud & Papi, a firm that belongs to Audemars Piguet.

It honors the brand A. Lange & Söhne and the people at work there that even in these times of modern design and production capacity the technical performances of watchmakers of past generations are also remembered. This includes among other things the movement's energy transmission by a fusee that is connected to the spring barrel by a fine chain.

Pour le Mérite

The Tourbograph Pour le Mérite is the first wristwatch ever to include a one-minute tourbillon, a chain and fusee, a split-seconds chronograph, and a power reserve display. This timepiece has a diameter of 41.2 mm and is 14.3 mm high.

The case back, embellished by embossing in relief, allows a view into the impressive movement, Caliber L903.0.

Both the tourbillon and the energy transmission of a movement by chain and fusee serve the goal of improving the rate precision of a watch. In order to reach this goal, two completely different paths were pursued here. With each of these brilliant mechanisms, the watchmaker is trying to compensate for different negative influences of the laws of physics on the rate of the movement—at different points within the movement.

The tourbillon is supposed to decrease the effect of gravity on a movement, while the chain and fusee serve

Pour le Mérite

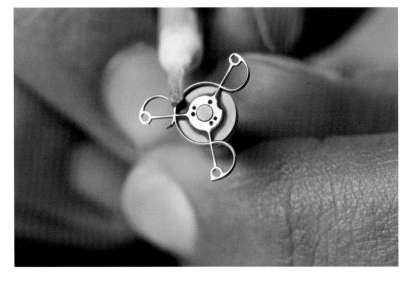

The regulation of the tourbillon takes place on a special mechanism that contains its own spring barrel before it is assembled along with the rest of the movement.

This filigreed component belonging to the tourbillon cage is secured to a holder in order to be worked by hand.

to utilize the mainspring's energy in an optimal way by only allowing a constant amount of torque to be passed on to the gear train. Both of these brilliant achievements in horology are based on comparatively simple concepts, but the technical realization of both is extremely difficult, though realized at A. Lange & Söhne with the perfection that the company is so famous for.

Chain Reaction under Tension

The chain-and-fusee transmission uses a clever mechanism to prohibit both the complete winding of a spring and its full relaxation. In addition, in this type of transmission the largest amount of the spring's energy affects the smallest action of the fusee (thus with the least amount of leverage), while the almost totally relaxed spring transmits its remaining power via the chain to the fusee (the greatest amount of leverage).

The design of this element of vintage technology in the Lange tourbillon Pour le Mérite from the mid-1990s and again in the Tourbograph Pour le Mérite introduced in 2005 goes a step further. Since transmission via chain and fusee has the fusee turning backward when it is wound, actually taking power from the movement during this time, a lavish, thirty-eight-component set of planetary gears was developed. This is placed within the fusee wheel, comprising a diameter of only 10 mm and ensuring that energy is transmitted to the movement uninterruptedly—even during winding.

The great advantage of consistency in the use of the mainspring's energy as a not-to-be-overestimated effect on the movement's rate is, for reasons of movement design, exchanged for the minimal power reserve of 36 hours.

After that a cleverly constructed lever simply closes off the energy, which is not a bad thing, but rather a necessity. Only this way can the tiny chain remain taut between the spring barrel and the fusee for another round of winding.

Caliber L903.0, a split-seconds chronograph driven by a chain and fusee and also featuring a power reserve display and tourbillon, comprises 465 individual components. Its plates and bridges are crafted in German silver, and the chronograph bridge is hand engraved. The 8.9 mm high movement has a diameter of 30 mm and is outfitted with 43 jewels, two of which are diamond endstones for the tourbillon.

Furthermore, upon pondering the fact that the tiny fusee must find room to house a steel chain 15 cm in length when the movement is fully wound, it becomes clear that a larger power reserve (that could only be achieved with the help of a longer chain) could not be realized for reasons of space alone—never mind the fact that a longer mainspring (including a correspondingly larger spring barrel) would also be necessary.

Only the Finest

Caliber L903.0 has a diameter of 30 mm and is 8.9 mm in height. The tourbillon makes its revolutions on the back of the movement underneath a cock and on the front underneath a graceful steel bridge, polished by hand for hours, that is secured to the dial by large screws located next to the numerals 4 and 8. These screws were very consciously taken into the design of the dial, as is the case with the Pour le Mérite tourbillon from 1994, on whose design the almost 3 mm larger Tourbograph is oriented.

Forty-three jewels function as bearings and endstones as well as pallets and the balance's jewel pin, ensuring decreased friction when the movement is at work. Six bearings are bedded in gold chatons that are secured to the base plate with little blued screws.

The flat parts of the new Caliber L903.0, like those of all Lange movements, are made of German silver, an alloy comprising copper, nickel, and zinc, which is used in its natural state without galvanic coatings. Before the plates, bridges, and cocks are assembled to form the movement, they have already gone through numerous work steps. Surface embellishment is one of the last of these.

While the front of the movement is dominated by the busily rotating tourbillon, the back seems to display a jumble of various bridges and cocks, springs, levers, screws, wheels, and two column wheels. All the steel parts of this movement are finely finished and beveled or are polished as a result of hours of work with a polishing file made of tin.

This movement, comprising 465 components, is housed in a platinum case (diameter 41.2 mm, height 14.3 mm) of which fifty-one pieces will be manufactured. Later in the ten-year production phase there will also be fifty pieces in other cases for a total edition of 101 pieces.

The design of the Tourbograph movement is based on that of the Pour le Mérite tourbillon, a member of 1994's debut collection (left). That timepiece is long sold out, only possibly available at prices for true connoisseurs at auction.

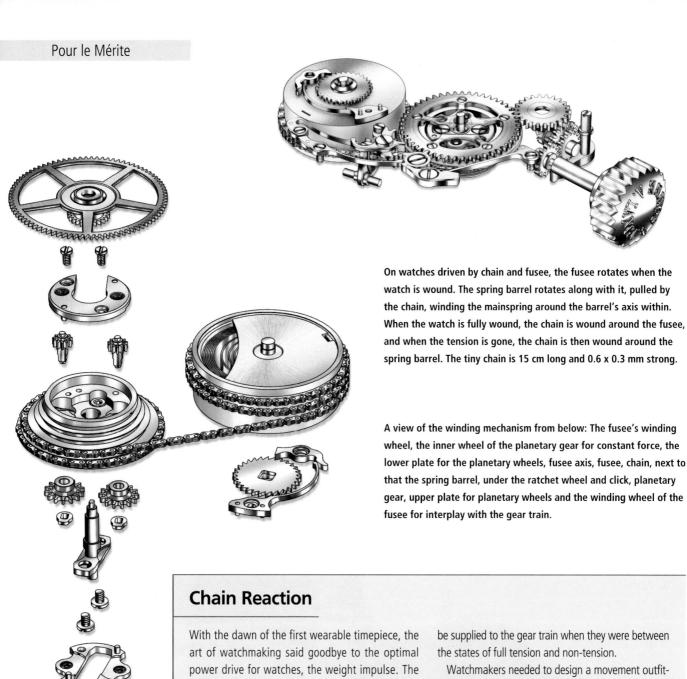

On watches driven by chain and fusee, the fusee rotates when the watch is wound. The spring barrel rotates along with it, pulled by the chain, winding the mainspring around the barrel's axis within. When the watch is fully wound, the chain is wound around the fusee, and when the tension is gone, the chain is then wound around the spring barrel. The tiny chain is 15 cm long and 0.6 x 0.3 mm strong.

A view of the winding mechanism from below: The fusee's winding wheel, the inner wheel of the planetary gear for constant force, the lower plate for the planetary wheels, fusee axis, fusee, chain, next to that the spring barrel, under the ratchet wheel and click, planetary gear, upper plate for planetary wheels and the winding wheel of the fusee for interplay with the gear train.

Chain Reaction

With the dawn of the first wearable timepiece, the art of watchmaking said goodbye to the optimal power drive for watches, the weight impulse. The advantage of mobility for a watch was traded out against the decisive disadvantage of uneven power impulses. A grandfather or wall clock driven by weight impulse is powered by continuously even energy. Improvements to a clock's rate precision are only necessary and possible in the gear train, escapement, and oscillation system (balance/pendulum).

Not so with a spring-driven watch. Here one of the causes for an imprecise rate could easily be the relaxing tension of the mainspring, a component that imparts its power not linearly, but in a curved manner.

The mainsprings formerly used were simple, flat steel strips that received the material elasticity they needed either by hardening and warming or by cold flattening. These springs had a hole on both ends. The spring was hung on a hook of the so-called core through one of the holes. The other end of the spring was hung on the barrel's hook, into which the spring was spirally wound. Because of the hardness and brittleness of the material, these springs caused varying amounts of torque to be supplied to the gear train when they were between the states of full tension and non-tension.

Watchmakers needed to design a movement outfitted with the balled-up power of a fully wound mainspring, but one that could also do its duty as precisely as an almost fully unwound spring. Thus they tried to utilize medium spring strength and constructed mechanisms to keep the fully wound and almost fully relaxed versions of the spring from affecting the movement.

The transmission of a watch movement by chain and fusee is, simply put, based on the law of the one-armed lever. To be more exact, it is based on the knowledge that the necessary leverage power decreases with the length of the lever used—in other words, if it is running out of power, all you have to do is lengthen the lever in order to get the desired leverage on the other end.

It is the latter that is utilized in this type of transmission by allowing the energy of the fully tensioned mainspring to be effective with the aid of the chain at the lowest amount of fusee leverage, and the almost fully unwound spring where there is the largest amount of leverage.

Pour le Mérite

Long a sought-after collector's piece: Lange's tourbillon Pour le Mérite ("for the merit") from 1994's debut collection. Platinum case, diameter 38.5 mm, height 10.5 mm.

Whirlwind in a Cage

The tourbillon was invented at the beginning of the nineteenth century by Abraham Louis Breguet, a man who blessed watch technology with numerous important innovations. With his invention of the tourbillon, Breguet attempted to compensate for the negative influence of gravity on the rate of a watch created by the unavoidable disequilibrium of the balance. Even if the balance is so exactly poised that its center of gravity is precisely in the middle (thus becoming ineffective), the setting of the balance spring alone will create new disequilibrium. Breguet's idea, ingeniously simple but technically difficult to realize, was to "move" the balance's center of gravity and have it render itself ineffective.

The tourbillon's so-called cage is perched upon the stem of the fourth wheel, where the escapement (pallets and escape wheel), balance, and oscillating system are hard at work. Since the cage revolves along with the stem of the fourth wheel around its own axis once a minute, any possible disequilibrium of the balance that may have occurred also rotates around once a minute, compensating for itself within the minute.

Since modern electronic measuring devices have been used for regulating watches, the disequilibrium of the entire oscillating system (balance and balance spring) can be detected in an assembled state and minimized.

The tourbillon, originally designed for use in pocket watches that were stored in vest pockets in a vertical position, has experienced a renaissance of late, but is basically technically unnecessary today. However, it is and remains the crowning glory of the art of watchmaking.

Watch Sensation of 1994

The tourbillon Pour le Mérite was introduced in the fall of 1994 together with the other debut collection models Lange 1, Arkade, and Saxonia. The name of the watch originates in the order inspired by Alexander von Humboldt and established by Frederick William IV in the year 1842 for outstanding performances above all in the area of natural science. At Lange the name was interpreted then as now to mean the performance of the doubtlessly outstanding talents of movement designers and watchmakers.

Connoisseurs admired the tourbillon, which back in 1994 was still a horological rarity, as well as perhaps even more the chain-and-fusee transmission, realized for the first time in a wristwatch's movement.

This highly complicated watch, released in a limited edition of 50 platinum and 150 gold pieces, was sold out at the end of the 1990s. ∎

TOURBOGRAPH Pour Le Mérite

Manufactured since	2005
Movement	mechanical with manual winding, Lange Caliber L903.0, diameter 30 mm, height 8.9 mm, 43 jewels, six in screw-mounted gold chatons, tourbillon cage bearings with two diamond endstones, shockproof glucydur screw balance, one-minute tourbillon, Nivarox balance spring, 21,600 vph; constant force regulation by chain and fusee as well as stepped planetary gear, power reserve 36 hours; Glashütte three-quarter plate; bridges with Glashütte ribbing
Functions	hours, minutes, split-seconds chronograph; power reserve display
Case	diameter 41.2 mm, height 14.3 mm, sapphire crystal, sapphire crystal case back secured by six screws
	limited edition of 51 pieces in platinum and 50 in other materials

Richard Lange

Simple Precision

Richard Lange is a timepiece commemorating the company founder's eldest son. With it, A. Lange & Söhne is celebrating the grand tradition of making especially precise observation watches, at the same time setting a counterpoint to the current trend toward ever more complicated watches.

The term "simple" is used in general language ambiguously and stands for both simple, as in modest, as well as for the reduction to that which is necessary, as in uncomplicated. As the word relates to watches, it could mean a modest or unassuming appearance and that the "uncomplicated" movement is oriented on the highest precision possible, this remaining the basic premise of timekeeping in the modern era.

With the new Richard Lange model, A. Lange & Söhne is reaching back to the tradition of (pocket) observation watches, of which there are many famous models from the era ending with World War II. These high-quality pocket watches, chiefly manufactured by hand, were already sought after as everyday watches for science and transportation in the second half of the nineteenth century due to their extraordinary rate precision. Above all, observation watches of this type were used on boats. There they served almost as a medium of transmitting the time between the highly precise ship chronometers located inside the boat, according to which they were set, and the officers on the command bridge or the afterdeck (a name in which the watch's other popular name is rooted: the deck watch), who were dependent upon the exact time for navigational purposes.

A large Lange observation watch from 1935. The little half-recessed button on the edge of the case above 11 o'clock needed to be pushed and held to set the hands.

 Richard Lange

Richard Lange

On the first German South Polar expedition from 1901 to 1903, the research ship Gauss had six precision pocket watches by A. Lange & Söhne on board.

Later, in the twenty years between 1917 and 1937, fifteen pieces of the Grand Observation Watch were manufactured in Glashütte, outfitted with a proud movement 57 mm in diameter and supplied to such reputable addresses as the Society for Timekeeping Science in Berlin, the Zeppelinwerft in Friedrichshafen, and the Physics Institute of the Mountain Academy Clausthal-Zellerfeld.

The size of the watches not only served the desire to make them supremely legible, something that needed to be guaranteed even under adverse light conditions; it also made for better and more precise setting, which was of course made easier by the large surface of the dial and the long hands.

The current trend toward larger men's watches oversteps here and there the boundaries of practical use. Today's watches are meant to be worn on the wrist and possibly even underneath a shirt cuff. The movement designers at Lange have therefore contained themselves to a case diameter of 40.5 mm. The bezel of the sapphire crystal on the case back found on all Lange watches is secured with six screws in the red gold, yellow gold, or platinum case.

A Movement of Special Quality

Richard Lange (1845–1932) understood like no other how to let scientific findings flow into his movement deliberations. He was especially interested in the balance spring as one of the decisive parts of the regulating organ. A scientific examination of nickel-beryllium alloys was published at the end of the 1920s, in which the compensating characteristic of a small addition of beryllium was described. Richard Lange was the first to recognize the use of these research results for the production of balance springs. He discovered that by mixing beryllium into nickel-steel alloys, the balance spring's sensitivity to temperature changes and magnetic influences decreased and at the same time its elasticity and hardness increased. In 1930 he registered the result of his research for a patent, calling it "Metal Alloy for Watch Springs."

As an homage to Richard Lange, the watch named for him is outfitted with a special movement: The manually wound Caliber L041.2, comprising 199 parts, with base plate, bridges, and cocks made of untreated German silver, was manufactured according to the highest quality criteria—decorated and assembled chiefly by hand and finely adjusted in five positions.

An additional set of wheels that turns underneath a bridge on back of the three-quarter plate maintains the rotation of the fourth wheel located outside of the movement (indirect sweep seconds). The movement can be stopped to precisely set the time.

Six Steps per Second

The large balance, working at 21,600 beats per hour, is regulated by turning the eccentric cams on its wheel; a balance spring buckle with index was not included. The outer end of the balance spring manufactured in-house is secured to a screw-mounted clamping mech-

RICHARD LANGE

Manufactured since	2006
Movement	mechanical with manual winding, Lange Caliber L041.2, diameter 30.4 mm, height 6 mm, 27 jewels, shockproof glucydur screw balance with eccentric regulating screws, in-house balance spring with spring clamp patent pending, 21,600 vph; power reserve 38 hours; Glashütte three-quarter plate; bridges with Glashütte ribbing
Functions	hours, minutes, sweep seconds
Case	diameter 40.5 mm, height 10.6 mm, sapphire crystal, sapphire crystal case back secured by six screws

The Richard Lange model in platinum and red gold. The long second hand made of tempered blue steel is beautifully combined with the gold hour and minute hands.

anism on the balance cock for which a patent has been registered.

The screw-mounted swan-neck fine adjustment found on the hand-engraved balance cock has only one function: to precisely set the so-called beat, or the position of the balance in relation to the escapement, by turning the finely threaded screw that belongs to it.

The solid silver dial is rhodium-plated on the platinum version of the Richard Lange. For use in the gold cases, the dial is silver-plated and printed with thin Roman numerals. The minute scale displays sixths of a second with the sweep second hand made of blued steel pointing to the partial lines located between each minute marker (21,600 vph or an oscillation frequency of 3 Hertz corresponds to three full or six half amplitudes). ∎

The Manufacture

Portrait Fabian Krone

The Electric Field between Tradition and Modernity

Portrait Fabian Krone

Fabian Krone came to Lange Uhren GmbH at the beginning of 2003, where he has been managing director with an operative focus on marketing, distribution, and finances. Since May 1, 2004 he has borne the responsibility that the position of CEO brings with it, sharing managerial duties with Hartmut Knothe, who is chiefly occupied with the company's production and technology and who coordinates the manufacture's extensive building projects.

A Paris-born management expert, who modestly calls himself a businessman, grew up in Peru and Italy and was previously with Italian automobile manufacturer Alfa Romeo for twelve years. There he was in charge of the company's worldwide distribution of sporty sedans and coupés. Working in a large concern is thus just as comfortable for the likable manager as dealing with the grand legacy of a highly emotional and traditional brand.

The technical background of the manufacture's history practically introduced itself to him. "Car people are always a little bit watch people too," says Krone from personal experience. He saw the renaissance of mechanical watchmaking up close in the 1990s as part of Italy's avant-garde society. "I threw myself into the world of A. Lange & Söhne's watches with the same passion that I committed to Italian cars fifteen years ago. And the great thing is that all the people in this brand's environment approach it with the same passion and enthusiasm—not only my colleagues and the company's employees, from the watchmakers to the engravers and bookkeepers, but also the dealers and, above all, the customers."

Krone was especially attracted to his new workplace by the opportunity to continue to build this technical and emotional brand, establishing it globally and positioning it clearly. On the one hand, A. Lange & Söhne makes modern watches outfitted with technical solutions that have never been seen before. On the other hand, its movements are outfitted with traditional swan-neck fine adjustment mechanisms, even though there are more modern solutions for regulating a movement's rate, and bearings set in gold chatons that are also no longer technically justified today. For Krone this is, however, not contradictory. In his eyes, A. Lange & Söhne's watches are "an incomparable amalgam of the rich tradition and the incredibly strong power of innovation that are present in this company. That's the electric field in which we live and from which we scoop our energy."

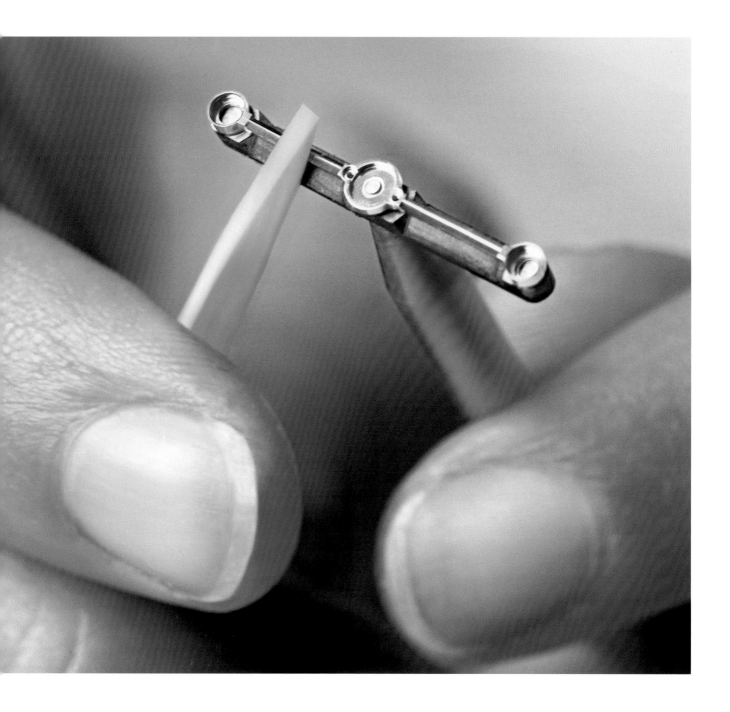

Straight on Course

During the last few years, a few new directions have been taken under Krone's lead—above all in terms of international distribution, which has been systematically broadened and at the same time carefully given a new structure, but also in the composing of the collection, which has been purified of reference details to make room for new, individual models.

The brand's committee for product development is made up of technicians, designers, and marketing people. "Thus, we can draw a complete picture of what we want to do in the next five years from different perspectives—what we need and what we can do," Krone explains with a grin. "The committee doesn't necessarily work in a democratic way, but the discussions are led with total openness. We meet about every two weeks."

The freedoms that the Saxon watch manufacture enjoys under the direction of the new owner, Richemont Luxury Group, are not written in stone, but must be worked for. "We set ourselves apart with interesting, well-thought-out, and doable ideas," Krone explains, "and we have earned the respect and trust of the concern's management throughout the last few years. We develop our products ourselves, though we clearly discuss them every year with the concern management.

"This is the correct way to go, for if someone has understood our brand, then he or she must also guaran-

tee that the core, the brand's DNA, must be maintained. We have all the freedom we need—naturally, within the frame of our possibilities. A brilliant situation."

Krone is responsible for the consistent maintenance of the brand's concept, the preservation of the spirit of Walter Lange and Günter Blümlein's vision, as well as a fine "adjustment" of the course that the brand will take in the future. Again and again he asks himself the question: "What is good for the brand? How can we maintain our tradition, the way the brand's founders would have seen it, without losing sight of our ties to the time we live in and the innovative energy of our brand? How would Ferdinand Adolph Lange have done it today?"

"The spirit of Günter Blümlein is still everywhere in this company, and the employees certainly get some of their creative energy from it. But it would detract from the achievements of our movement designers and watchmakers if we were to say that they were only developing fifteen-year-old concepts. We have nothing more left in the pipeline from back then. Even after the era of Blümlein and Meis came to a close, our employees have created fantastic concepts that have already been realized or will be in the future."

"... we set ourselves apart with interesting, well-thought-out, and doable ideas ..."

Building on the Future

What Lange earns on the sale of its watches is for the most part reinvested. "We put a lot of money into new technology, into the development of new products and the training of our employees, but in the end a little bit does remain," Krone says with a grin.

The permanent personnel and spatial expansion of Lange Uhren GmbH (the old Glashütte brewery on the other side of the street has meanwhile also been purchased) serve not to increase production numbers, but to secure quality: A. Lange & Söhne's watches are not only more complicated and refined, but the number of parts and subgroups manufactured in-house is also continuously increasing. "In 2005 alone we hired fifty-four new employees," says Krone, "and every year we wait impatiently for the graduates of our school of watchmaking." ∎

The Manufacture

Heartfelt Handwork

In the end, any company is only as good as its employees. This is even truer of watch manufactures, where products are created in long labor-intensive processes. A. Lange & Söhne does own a set of highly modern machines to do some of the work, but the most important bits are and will remain the traditional work done by hand during production, finishing, and assembly.

The Lange manufacture has continuously expanded in the last fifteen years, and the amount of space it takes up is considerable. There are now more than 400 employees, making this the largest company in the city of Glashütte, spread over five buildings along Altenberger Strasse at this point in time.

The main historical factory building located at the junction in front of the train station has once again become the spiritual center of the company. This is where the memories of the Lange family's traditions reside, this is where—freshly refurbished—the large pendulum clock that for many years set the beat of Glashütte's time once again ticks. After Lange repurchased it from the city of Glashütte in 2001, the building was completely and lavishly renovated, shining inside and out with a new glow again today. At the address Ferdinand-Adolph-Lange-Platz 1, distribution, event marketing, and bookkeeping find their home. And until the new apprentice workshops in the large complex of buildings on the grounds of Glashütte's former brewery are completed, the apprentices of the company's own watchmaker school follow in the footsteps of historical watchmakers here as well. Their workshops are located in the former production rooms in the west wing of the main historical factory building together with the workshops for the engravers and case restorers.

While the roots of this traditional company are in the main historical factory building, the nucleus of Lange, refounded after the fall of the Berlin Wall, is moving piece by piece up the street in the direction of Altenberg. The building internally known as Lange I once belonged to the precision clock factory Strasser & Rohde, though it was never used for production. That changed directly after its renovation. Watchmaker benches, small lathes, and various other mechanisms were installed to assemble the first Lange watches of the modern era. Today, pre- and final assembly of the complicated movements are still found here as are the development and prototype departments and a special workshop for "inventing."

On the same side of the street, separated by a cobblestone square, Lange II is found in the former Archimedes adding machine factory. This was outfitted with modern toolmaking machines, work centers, and measuring technology during mid-1990s and is now home to the company's raw component production—though only on the ground floor and mezzanine levels. The upper floors are home to a showroom outfitted with museum pieces, a modern reception area, and generous sitting and training rooms as well as the management offices.

Behind Lange II, a long, low, stretched-out edifice parallel to the main building can be seen, christened Lange III. It is here that one finds various workshops for deburring and decoration. Also in the backyard, so to speak, but closer to Lange I and actually joined to it by a covered bridge connecting the second floors, a squared glass building is to be seen: the Technology and Development Center (TEZ). Due to its connection with Lange I (similar to that found between the Dresden Castle and its court church and good for going back and forth in the winter to keep one's feet dry) and, of course, because of the practical connection to the assembly workshops, the glass building is called Lange IA. Here is where A. Lange & Söhne's research and development center as well as quality control department are found, where the logistics planners have their offices, where new technologies are thought up and tried out, and where one of the most modern testing laboratories in the entire watch industry is located. Furthermore, here is where the filigreed little balance springs are wound and the highly complicated watch movements (chronograph upwards) are assembled.

The Manufacture

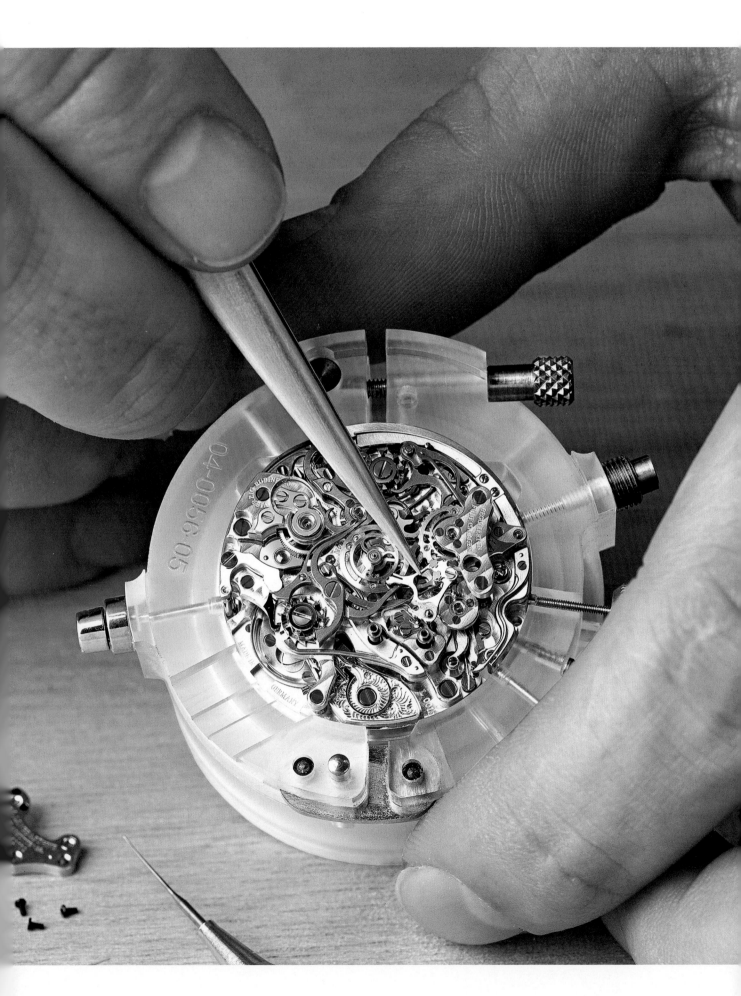

A. Lange & Söhne – the Employees

Tino Bobe, the manufacture's manager.

Anthony de Haas, director of product development.

Manufacture Management

At Lange IA, Tino Bobe, the manufacture's director, keeps his office. His workplace is actually to be found throughout the company, in all five buildings, for strictly speaking there should not be anything going on in the manufacture that he doesn't know about. Bobe is constantly in contact with each departmental head, who informs him of the current production status in his or her area of responsibility. "Above all, I am a popular person to talk to when something doesn't go the way it was planned," Bobe laughs. "When everything functions properly, and all goals are fulfilled and the work is done with no friction, then all are happy—and quiet." Then he can take care of the organization of the next tasks at hand—in the hope that everyone remains quiet.

Bobe is the interface between marketing needs, developmental possibilities, and manufacturing capacity, representing an important connecting link between distribution, product design, movement design, and production. As a movement designer who has made his mark and longtime project manager, he was perfectly prepared for this very responsible position.

Design, Development, Movement Design, and Prototypes

New Lange watches are first created in the heads of the movement and watch designers, with a fairly balanced relationship between the two. Some watches are first devised by their functions, which need to be "packed up" properly, while others come to life by someone having an exciting design idea that needs to be completed with a function.

The small group of creative heads, Martin Schetter, Ilka Smyreck, and Nils Bode, have occupied themselves intensely with the characteristic elements of the collections and create every new model design with the fine feel they have developed over many years in order to give Lange's philosophy concrete form. As during the time of Günter Blümlein, a man who heavily contributed to the appearance of A. Lange & Söhne's watches, these designers still put emphasis on the fact that the watches should have a strong but restrained appearance. With the Tourbograph Pour le Mérite model they succeeded in creating a true homage to the Lange renaissance initiator four years after his premature death.

Movement designers Helmut Geyer, Karin Dinter, Burkhard Geyer, Axel Bobe, and Annegret Fleischer.

Anthony de Haas, director of product development, meets with his teams regularly to exchange thoughts. He has managed this department since 2004, having previously worked in similar positions at IWC and then Renaud & Papi, the reputable Swiss specialist for complicated watch movements. To develop movements for Lange's watches, it is not only his great deal of experience in the area of movement design that is invaluable, but also his thorough knowledge of the watch market in general.

The movement design team at A. Lange & Söhne has grown dramatically in the last fifteen years in order to properly execute the exponentially expanding number of projects. Annegret Fleischer and Helmut Geyer have been with the company practically since the day of its refounding. They were previously employed together at VEB (Company of the People) Glashütter Uhrenbetrieb and without a doubt stamped the first Lange wristwatches of the modern era with their signatures. Jens Schneider came from the prototype department over to the movement design team as a specialist for technical troubleshooting in production due to his practical experience. Draftsperson Karin Dinter and movement designer Axel Bobe are two more reputable experts completing the team. Geyer is already toying with the idea of vacating his seat after the completion of the next big project and passing on the direction of this department in order to dedicate himself entirely to the movement projects he has planned. Geyer's son Burkhard joined the squad in 2002, and the father-son team is the best example of how through joint daily work the know-how of one generation is passed on to the next according to Glashütte tradition. Burkhard Geyer was one of the first two Lange apprentices of the modern era, completing his studies with excellent grades as illustrated by a complicated pocket watch including an equation of time display that he made.

Designers Nils Bode, Ilka Smyreck, and Martin Schetter.

The Manufacture

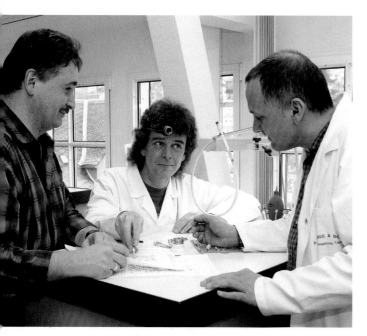

Movement designer Jens Schneider with prototype makers Thomas Gemmel and Andreas Gelbrich.

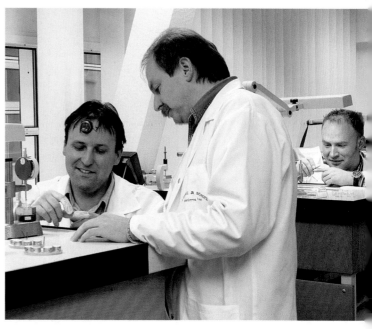

Thomas Weiß, Udo Rudolf, and Carsten Scheppler of the prototype department.

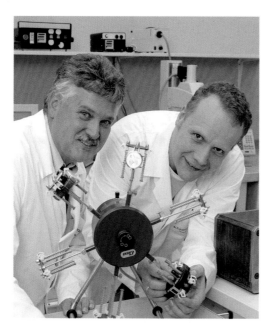

Bernd Glöckner, quality assurance, and Christoph Schlencker, testing laboratory.

The movement design department works hand-in-hand with the prototype department—this also being a generations-old Glashütte tradition. The most resourceful inventors in the country are at work here, and not rarely do difficult detail solutions become obvious as they work concretely on the material object—where it can come to light that some tolerances calculated by the computer are really not enough "in metal." Movements are now designed every day on the computer with the aid of corresponding programs that allow a myriad of motion simulations for individual components on-screen. The prototype is thus a relatively late step in the movement design game, with the tasks of Udo Rudolf, Thomas Gemmel, Carsten Scheppler, Thomas Weiß, and Andreas Gelbrich even more detailed and filigreed. Naturally, the all-round watchmakers of the prototype department are also popular experts for special jobs such as the restoration of the Gutkaes pendulum clock in the main Lange factory building and the famous astronomical grandfather clock by Hermann Goertz that was completed under Gelbrich's direction.

Testing Laboratory

Before the big machines of the production area can be turned on for component production of a new movement, individual movement mechanisms and even complete movement prototypes must undergo testing in Christoph Schlencker's "torture chamber". His facilities make up one of the most modern laboratories in the watch industry and can simulate every conceivable pressure and strain (continuous function, shock, heat, cold, and vibration tests) that a watch movement may expect to find out there in the wide world. The same laboratory facilities are also used in the Technology and Develop-

ment Center (Lange IA) by Bernd Glöckner, head of the quality assurance department, a man who has collected a great deal of experience in the horological world as a watchmaker, precision engineering technician, and movement designer—a fact that is a distinct advantage to him today in the independent and objective analysis of completed products.

Balance Spring Development and Production

Lutz Grossmann and Rainer Kocarek also work under laboratory conditions, though consciously outside the strict cycle of the production area, on their "black arts," aka balance spring production. Lange is one of the very few watch manufactures in the world able to make its own balance springs. Until now it has been top models such as the Double Split chronograph and the Datograph Perpetual that have benefited from the in-house balance springs.

The two engineers were already making balance springs in the GDR era and have in the meantime perfected the process of drawing, rolling, coiling, and treating the paper-thin spring wire five hundredths of a

Rainer Kocarek and Lutz Grossmann in balance spring production.

millimeter thick with heat. Grossmann busies himself with the theory of material composition and shaping, while Kocarek is an ingenious mechanic with a special feel for the adjustment and operation of machines and tools. Especially noteworthy is the fact that the pair has developed a measuring and control technology that is necessary to the entire process.

The Manufacture

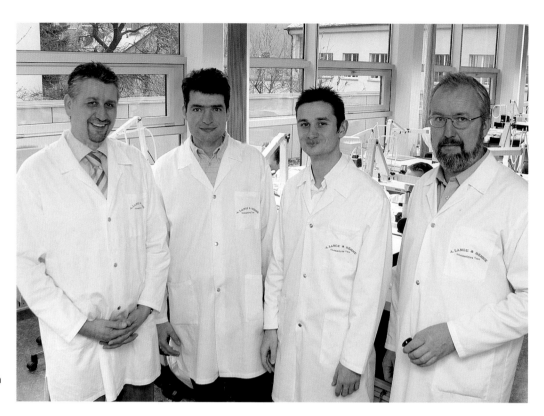

Bert Salomon, Sven Richter, Mathias Zschiedrich, and Michael Sachse of production planning and control.

After all the departments portrayed until now have attended to the basics of conception, development, and movement design, it's slowly but surely time to head to the production of innumerable individual components. Here, as well, it's not so much the number of pieces produced that make these employees' jobs harder, but rather the many different individual parts that total in the thousands within the pool needed for twenty-three independent calibers with several hundred individual components per movement.

Production Management

Bert Salomon is responsible for the entire process. He generates the tasks for the production department, has additional people hired, or moves them to different departments where they are needed more at a given moment. He must determine whether the given tasks and activities can be managed with the workforce at his disposal. He is assisted by Michael Sachse, one of the company's first employees, and Sven Richter, who steers production, and also works closely with Mathias Zschiedrich, who is responsible for production planning.

New employees must first go through a series of different departments before they can be utilized in the area for which they were originally hired. Such "learning curves" can last up to a year, and these delays are some-thing that Salomon and Zschiedrich must calculate into the mix—not an easy job, especially since qualified people are hard to find.

Toolmaking

Toolmaking lies under the direction of Franz Rudolf, who, like the majority of his colleagues in the department, is a mechanical engineer and was employed making the movements' chassis, better known as plates, for many years. He has actually reached retirement age, but in his more than ten years at Lange, he has shaped the toolmaking department into an efficient partner for the production departments. Custom mounts, mechanisms, drills, millers, and other tools as well as smaller machines are created in his department—if need be, even overnight.

Component Manufacture

The first shavings actually fly on the ground floor and in the basement of Lange II. The big work centers are located here—computer-controlled tooling machines the size of minivans that make the most filigreed objects from unshaped blocks of brass, German silver, steel, gold, and platinum under a never-ending stream of cooling lubrication liquid. In two shifts, plates, cocks, and

The Manufacture

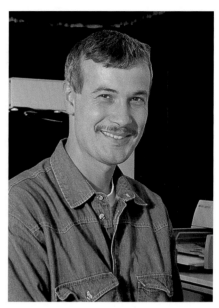

Franz Rudolf, director of toolmaking, and Dirk Rauchfuß, head of parts manufacture.

bridges are milled, while steel springs and levers are eroded by the wire spark method.

Toolmaker and master precision engineer Dirk Rauchfuß is the head of the production center. His area of responsibility is also becoming more and more diverse with the increasing number of more complex movements in the collection, corresponding to the new models that are continually being introduced. The effort made to coordinate and continuously maintain projects is thus ever more demanding.

Michael Morawietz and Steffen Rinke are the shift foremen of the "blue boys"—so called because of their laid-back jeans-style work clothes. In the measuring and milling center, Benito Lukas directs the wire spark erosion facilities in the basement, where complex and filigreed little parts are cut out of a metal plate that has already been worked by CNC technology on the most modern of machines using a laser-sharp ray of light and a hair-thin brass wire under water or oil. Tino Weller guards over the automatic long lathes located in the turning and teeth-cutting department, where pinions, wheels, pegs, stems, and screws are made, now at home in TEZ's basement (Lange IA).

From left to right: Steffen Rinke and Michael Morawietz from the measuring center; Benito Lukas, in charge of wire spark erosion, and Tino Weller of the turning department.

127

The Manufacture

Margit Nitsche (second from left) and her team are in charge of deburring.

A view of watchmakers at their benches in the finishing department.

Deburring and Finishing

Before the individual components of the movements go to the various assembly departments, they are inspected with a loupe by experienced specialists. The deburring department on the second floor of Lange III examines the parts coming from production and brushes them up by hand—some even before they go to the wire spark erosion machine for more work, for example, in order to ensure that they sit correctly in their holders. Every edge of every hole, cutaway, drilling, and free surface is examined and possible burrs trimmed. This work demands as much care as it does a feel for the material and is thus a strong female domain in the world of watchmaking. With special sharp, pointy tools, reminding one of little knives, scrapers, spatulas, and spoons, these specialists trim away, preparing the parts for further processing.

Many components go from deburring directly to the various finishing departments where they undergo different surface treatments depending on the material, function, use, and positioning of the part. Department head Kai Tschochner and "operative technician" Maik Pfeifer discuss how the demanding processes that need to be applied to the various individual parts can be completed by the more than forty men and women here.

Together with Stefan Graubner from the "technology" department, Pfeifer, who actually built up the finishing department, thinks up different decorations, perlages, bevelings, and polishings, setting down a certain process for each and every movement component.

At Lange III, decorations and perlages are applied to plates, bridges, and cocks with various small table machines driven by electric motors and thin drive belts.

Fine surfaces and perfect edges: Clean finish is everything to these fine watchmakers. Christine Marks, Marlies Fraulob, and Marita Grän.

Kai Tschochner and Maik Pfeifer head up the finishing technology.

A pattern of up to fourteen different perlage diameters decorates the plates of a movement, not only on the back, but also on the dial side, where no one will probably ever get to see it. For every pattern there is an individual rotating stamp or rubber peg, which is pressed down for a short amount of time against the surface of the plate to be decorated. After that, the piece is turned about half the diameter of one of the pearls, and another one is set down right next to it—and so on and so forth, pearl for pearl. When the row is done, a second row of concentric pearls follows. Some perlage machines have an automatic rotating chuck that makes the precise placement of the pearls possible. But it goes quicker freehand, if the experience is there and the routine established.

Glashütte ribbing is applied using a half-automatic mechanism. The machine moves a polishing head that is not quite vertically positioned (similar to the perlage peg) in parallel lines across the piece, creating stripes that charm a discreetly iridescent play of light onto a movement's surface.

Beveling, stripe polishing, and mirror polishing are, however, processes completed entirely by hand, which not only demand a great deal of feel for the material, but a great deal of patience as well. Beveling is the process of minutely filing the edges of flat parts and levers at a

The Manufacture

Elke Weller, perlage

Hannelore Tänzer, hand finishing

Regina Bachmann, circular finishing on wheels

precisely determined angle, with the surface created on all the edges being kept the same width. Stripe polishing creates what appears at first glance to be a raw or nonworked surface, but what is in reality the result of a very precise and even polishing motion made with a special file. Mirror polishing is pretty much the exact opposite of stripe polishing and is created by rubbing a part unwaveringly and consistently on a tin plate. When the surface is absolutely flat and smooth, it reflects light rays only in its angle of incidence.

Frank Wolf, director of the assembly department.

Watch Assembly

Frank Wolf is in charge of the assembly department and thus responsible for more than eighty highly specialized watchmakers. He has spent his entire professional career in this core area of watchmaking and shows his special qualities in assembling balance springs and the fine adjustment and regulation of complicated watch movements.

A movement, especially a complicated one, is created in several stages, going through various watchmaker departments. All Lange calibers are put together at the end from preassembled subgroups, and thus there is a concentrated troupe that specializes in preparing components.

Especially delicate is the preassembly of the escapement, for at A. Lange & Söhne all balance springs, balances, and pallets are still put together by hand. Waltraud Siegert directs this department, assembles the pallets, and completes the oscillating system. Angela Weber prepares the balance springs and bends the terminal curves individually for each watch and, it goes without saying, differently for every caliber. Martin Jonas levels and centers the balance springs.

Raya Rauchfuß supervises the preassembly of individual subgroups such as the addition of the swan-neck spring to the balance cock, while Elke Göhler and Ursula Göhlert show watchmakers how to put together gear trains and add the winding system, all the while making sure that the endshakes are precisely regulated.

Irka Zeibig's team puts the large date comprising more than sixty individual components together, including its complex changing of the single and double digit disks. Some of these very specialized watchmakers are also responsible for the preassembly of the perpetual calendar's complicated mechanics, which can take one watchmaker several working days.

In a difficult process, Angelika Weiß and her fine adjusters regulate watches in the traditional five posi-

The Manufacture

Base movements and complex subgroups such as the outsize date and the escapement are preassembled by expert hands: watchmakers Ursula Göhlert, Angelika Weiß, Irka Zeibig, Elke Göhler, and Raya Rauchfuß.

Waltraud Siegert, Martin Jonas, and Angela Weber assemble escapements.

Jan Helbig is a watchmaker in the chronograph assembly department.

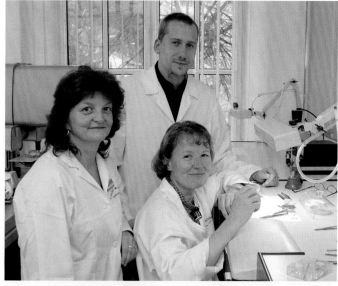

tions until only an extremely small daily rate deviation is reached. This process takes place during first assembly and once more after the watch has been completely assembled.

The chronograph makers under the direction of Andreas Müller fit the components they have received to the proper dimensions, meaning they file and grind them by hand until the chronograph levers and catches work perfectly in tune with each other, transmitting all energy and functioning with as little friction as possible. After that, they add the completed chronograph module to the base movement, where further nitpicky adjustment work is done. The really complicated movements such as the Lange Double Split, the Tourbograph Pour le Mérite, and the perpetual calendars are assembled by a handful of experts in a difficult process practically as independent pieces.

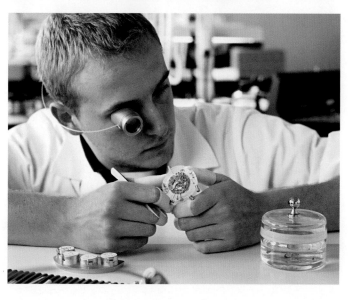

The Manufacture

Kerstin Richter, final assembly.

Andreas Müller, chronograph assembly; Marina Göhler, head of final control; Andrea Glöditzsch, head of encasing.

The movements' final assembly is entrusted to the especially versatile among the company's talented watchmakers. In Kerstin Richter's team, every movement is again taken apart, cleaned, lubricated, and assembled a second time on the finished three-quarter plate. This has nothing remotely to do with serial production, for it is the purest form of manufacture culture as it was practiced 150 years ago.

In Andrea Glöditzsch's department, completed watch movements are housed. This means the dial and hands are placed upon the movement, which is then put inside the case and screwed to the sapphire crystal case back. Marina Göhler and her team are responsible for the watch's final checks, and they examine every nook and cranny of every watch. The movements remain on the movement simulator for several weeks—simply called "the mill" at Lange—so that they can be checked against a radio-controlled clock every day.

Old Watchmaking Traditions

In the west wing of the main Lange building, where about one hundred years ago the production of the famous pocket watches was housed, tradition is at home today—in the form of a small preserve of those arts done by hand that are in danger of extinction.

Here is where master engraver Helmut Wagner works with a handful of young artists who adorn each and every balance cock of a Lange watch under the microscope with their individual signatures and additionally fulfill almost every client's special desire. This is where production and education intermingle as in the olden days, and learning is done in "real time" on the object.

Next door, an interesting two-man team has taken up residence in this little niche: Rainer Pellmann and Manuel Schneider create highly fine pocket watch cases using only archaic tools and their hands. Pellmann, the older of the two, brought the younger Schneider to this craft, which might seem like a mixture of forge and mechanical engineering workshop. Meanwhile, Schneider is a "young master" and understands much of the preparation and manufacturing of these cases, which comprise

True artists with a lot of feel for the material:
engraver Helmut Wagner and his team.

Master watchmaker Peter Lötzsch during final assembly of a movement.

several parts including hinges. He can't compare his to Pellmann's experience, who shortly after the fall of the Berlin Wall managed his own workshop for pocket watch restoration together with Jan Sliva. Pellmann and Sliva followed the call of A. Lange & Söhne together and moved their workshop to Glashütte.

Jan Sliva is Lange's man responsible for historical watches that customers or museums send in for maintenance or repair. Naturally he works hand-in-hand with his former business partner. The most damage he has ever had to right at A. Lange & Söhne was caused by the devastating flood of August 2002, which affected the company the most in the basement safes of the main factory building. The watches that were stored there—partially restored already—were pocket watches from private customers, which then needed to be completely taken apart, dried, cleaned, lubricated, and reassembled in the ensuing phase. This was a task absolutely worthy of Sisyphus for Sliva, who prefers to work alone.

Jan Sliva, workshop for historical timepieces.

Manuel Schneider and Rainer Pellmann, pocket watch cases.

The Manufacture

Head apprentice instructor Katja König and instructor Sebastian Rentzsch surrounded by Lange's apprentices.

Apprentice Education

Lange's school of watchmaking, as previously mentioned, is also located in the main factory building, in light, glass-encased rooms, outfitted with sliding glass doors to the main hallway.

Katja König and Sebastian Rentzsch always have two classes under their watchful eyes, while a third enjoys classes in the vocational school center in Glashütte. When that school's on break and all eighteen to twenty apprentices are present in the Lange school of watchmaking at the same time, it's a tight fit, but everyone finds room.

It's far more difficult to even get an apprenticeship at Lange, as head of the company's educational department König explains, "From more than a hundred candidates who apply each year, we only have room for twelve to fourteen of them. Many come from the surrounding region, but also from a variety of different countries." Watchmaker education at Lange is based on a general teaching plan, but the Glashütte theme does enjoy special status. The tradition begun by the German School of Watchmaking remains in the conscious part of the brain, as well as the company's own contributions to the history of timekeeping, of course.

In the first year of apprenticeship, basic skills such as filing, sawing, and turning are conveyed and tested on examples of clock movements. The goal is to be able to complete simple repairs, make parts according to patterns, and create one's own tools, which will at least accompany the junior watchmakers throughout their apprenticeship. In the second year, they begin with pocket watches and the theory of portable timekeeping, including the balance and lever escapement, while in the third year, filigreed wristwatches are finally brought to the bench.

They practice on anything that ticks—the apprentices may bring watches from home to repair, and there is also a fund of old and new watch movements made by various manufacturers.

Preparing them for the reality of Lange's modern watches, König and Rentzsch send their charges off to internships in the various production departments. In the first apprentice year, they already go for a few days to the trimming and decoration departments, and later to preassembly to breathe a bit of "manufacture air," meet colleagues, and begin to decide for themselves which of these activities could be the one they can do best. Most of the apprentices stay at A. Lange & Söhne and may already know on the day they graduate in which department they would like to begin their careers.

Customer Service and Service-Marketing

Norbert Windecker was also once an instructor in Lange's school of watchmaking, but the born Hessian

The Manufacture

has meanwhile changed fronts. After three and a half years in the educational workshops, he now manages the service workshop for the manufacture and, with a team of nine watchmakers, takes care of all kinds of repairs and revisions to Lange watches. Windecker's workshop deals with about 60 percent of all the repairs necessary worldwide. This is where the specialists for complications such as tourbillons or a Langematik Perpetual are located. Service work on highly complicated calibers and repairs to strongly damaged watches are only done in Glashütte.

Lange has external service centers in Schaffhausen (Switzerland), New York, Tokyo, and Hong Kong, where German master watchmaker Manfred Weber does more than just run a workshop; he personally takes care of his local customers. "In China, I am more of a local Lange representative than anything else, a type of ambassador, if you will. The dealers, the consumers, and especially the collectors, want to speak with a real watchmaker in order to get technical information firsthand."

At Lange customer service is still understood as a service to the customer, and it is a correspondingly high priority. For this reason, the Service- Marketing department was called to life two years ago, under whose roof technical customer service and the care of customers is united under the management of Benjamin Lange, Walter Lange's son. Benjamin Lange is a creditable representative of the "family firm," as Lange, despite being embedded in a larger concern, is still perceived today. He travels a lot to spread the Lange company values around to concessionaries and customers.

Gudrun Pahl has worked for Lange for more than nine years and is today in charge of caring for customers. Alongside answering queries from all over the world and performing service administration for current watch models, she and her team also work with historical Lange watches—at Lange tradition is carefully preserved.

Occasional inevitable complaints by customers usually concern the necessity of and length of time needed for overhauls and repairs. In these cases, the deparment's friendly colleagues patiently explain the necessity and extent of an overhaul, which is recommended every three to five years. Most of the time, these customers have not thought about the fact that a watch by A. Lange & Söhne, depending upon the model, can have up to 500 components working together—twenty-four hours a day, 8,760 hours per year. The fact that such a miniature high-performance mechanism undergoes wear and tear is just a matter of course.

Norbert Windecker, head of the service workshop.

Benjamin Lange, Service-Marketing manager, in discussion with Manfred Weber of Lange's service center in Hong Kong.

Sandra Knorr, Christel Alpermann, Gudrun Pahl, and Claudia Sawodni, customer care.

The Manufacture

Sebastian Vossmann, distribution director.

Joanna Gribben, Lange Academy.

Manuela Wolf, shipping.

Jenny Mädel, Jan Abele, Yvonne Lasch, and Ellen Reimann of Event Marketing.

Distribution

Sebastian Vossmann, a business manager from Hamburg, heads up Lange's worldwide distribution, working with the brand managers in the various markets, those men and women who have contact with the local concessionaries. Through this, Vossmann is able to analyze the markets in terms of customer needs.

The director of distribution is also responsible for his colleagues in Glashütte, who receive orders from authorized dealers in more than forty countries around the world every day, making reservations, writing invoices, and working with complicated duty formalities.

Manuela Wolf, who is in charge of shipping, also belongs to the distribution team. Every day, she makes out guarantee labels, delivery memos, and shipping contracts for dozens and dozens of watches—timepieces that the watchmakers have finished assembling and checking, finally placing them on her desk.

Organizationally speaking, the new Lange Academy belongs to the distribution department, for what Joanna Gribben is building up here is no less than an intelligent sales promotion instrument. "Creating an awareness for the brand," is how the Irish national describes the program for schooling sales personnel, which is divided into different courses. "Understanding Lange" establishes the basics for occupying oneself with the thoughts of the manufacture tradition and its positioning on the market. "Living Lange" is supposed to create a relationship with people who have contact to these wares every day, teaching the philosophy of the A. Lange & Söhne brand in seminars and workshops in Glashütte.

Jan Abele directs Event Marketing, a department that has occupied itself for years with the conception and realization of customer events and trips, exhibitions, and trade shows. This department is well filled with creative organizational talents, which is why Yvonne Lasch, Jenny Mädel, and Ellen Reimann very often travel, ensuring the

Lange brand managers for individual export markets.

company's high demand on quality at every local event. It is personal contact with customers that often especially contributes to the positive development of the A. Lange & Söhne brand.

Marketing

Right from the beginning, an important building block in the successful concept of the A. Lange & Söhne brand has been the well-organized and creative marketing and communication department. Annette Bamert was literally involved in the A. Lange & Söhne project from the very first minute and is responsible for all marketing activities from the conception of advertising motifs to the design of catalogues, brochures, and price lists and the realization of film and book projects on the topic of A. Lange & Söhne. She has a powerful marketing and communication team at her side.

Annette Bamert, marketing director.

Public Relations

The PR department under the direction of Arnd Einhorn profits from its proximity to the manufacture. Important information can be gathered directly for representatives of the international press or private watch aficionados. The organization of press events all over the world belongs just as much to their daily routine as keeping the company archives, analysis of press clippings, and the competent care of numerous queries by watch collectors via the Lange website.

During a tour of the manufacture, PR representative Enver Nickel patiently and in great detail explains everything imaginable about the creation of a Lange watch. And four times a year he also creates the *Lange-Anzeiger*, an internal company magazine for the employees of Lange. ∎

Arnd Einhorn, head of the PR department.

Portrait Hartmut Knothe

Right from the Beginning

Portrait Hartmut Knothe

The Lange historic family domain just after the fall of the Berlin Wall (bottom) and after extensive renovations (left).

... factory manager with extensive special duties ...

Hartmut Knothe, one of the managing directors of Lange Uhren GmbH, is one of those few men who were there right from the beginning. A toolmaker and engineer by trade, he had earned three additional correspondence degrees in various technical and pedagogic disciplines by the age of thirty-eight. As the longtime director of the trade school in Glashütte and a member of GUB's managing board of directors, he had the best prerequisites for putting together the core workforce of Lange Uhren GmbH. At first he was titled "factory manager with extensive special duties." These special duties included among others overseeing building construction in Glashütte, installing the tooling and production machines, organizing management and production structures, and communicating with the Treuhand and state organizations.

As a specialist for watchmaker education in Glashütte, he was very familiar with each of the individual specialized departments, their technical and personnel structures, and their most talented craftsmen. "I profited from this trusting relationship later when it came time to recruit employees for Lange," he remembers.

At the beginning, Knothe was still contractually bound to GUB's training program. However, at the time of German reunification representatives of the Treuhand became aware of this man who, during the uncertainty of those days, was very popular due to a good set of feelers put out in all directions. He was an important information carrier and contact person. After working hard at restructuring the "people's company" and turning it into a western-style limited company, Knothe was offered the position of managing director of the somewhat disoriented Glashütter Uhrenbetrieb.

In the end, Knothe declined the offered job. The Treuhand plan called for the continued employment of all 2,000 GUB employees, and Knothe was already convinced that previous production and personnel structures urgently needed to be revised in order to survive the new economic situation.

Knothe was in fact already working on plans for a private education center when—as fate would have it—he met Walter Lange and Günter Blümlein in the waiting room of the mayor's office on one of their first visits to Glashütte. They introduced themselves and learned that they had already heard of each other. They made an appointment to talk—a meeting at Glashütte's Ladenmühle guest house that later went down in Lange's history.

The renovations on the production buildings located on Altenberger Straße lasted many months. Today the buildings, named Lange I and Lange II, make up the characteristic appearance of the street.

Everything came down to word-of-mouth, a grand old name ...

A Few Days to Think It Over

Knothe was not immediately excited about the ambitious ideas of the manager and the nostalgic dreams of the native son. He asked for some time to think it through, analyze his commitment to the project, and discuss it with his family. Then he took an adventurous trip to see Günter Blümlein in Schaffhausen, where the planning office for "Project Lange" was set up on IWC's upper floors. He had brought ten sheets of paper full of questions in his briefcase that took an entire day to work through. The quality of these questions and the quality of the answers were the key to the unusual friendship that was to develop between the two men from a divided Germany. Knothe decided to leave GUB.

In the ensuing period, the engineer began to take on more of a role as intermediary agent alongside his logistical duties. It was obvious that the watchmakers in Glashütte were very interested in the new project but they were still quite ambivalent. GUB was having difficulties, and although its future was uncertain, it was still in existence. There was no tangible evidence to prove that Lange's company would be realized. Everything came down to word-of-mouth, a grand old name, and the grandiose promises of a West German.

"Günter Blümlein was very reserved with promises at the beginning," Knothe remembers, "we made a virtue of it, even later only promising as much as we were sure we could keep—then delivering more than was promised."

While Knothe worked on searching for a fitting building, negotiating with the government about restrictions and making initial renovation plans, he continually received visitors in the form of former apprentices and employees. They came to his temporary office at Lange's headquarters and asked him his thoughts on the situation. Was Lange Uhren GmbH a respectable company? Would watches really be manufactured here again? And why not at GUB, where everything was already set up?

Just a few weeks prior, he had asked Blümlein those very same questions, and now it was up to Knothe to answer them in the same convincing manner. "I learned a lot from Günter Blümlein and view each one of those eleven years in which I was fortunate enough to work with him as a personal gain," Knothe says of the turbulent nineties.

Portrait Hartmut Knothe

Rebuilding East Germany

He shrugs off his own contribution to the success of the project, "I was maybe important as an intermediary between the two industrial cultures, the two mentalities, the two systems, if you will. In the end, many of our first employees certainly ventured over to Lange because I was so obviously convinced. People said, 'If Knothe has gone there, it can't be all bad. He has always had a good nose.' Without an East German as an intermediary, the first contacts would, without a doubt, have been much, much harder."

In all his modesty Knothe overlooks the fact that he was in charge of creating the company's technical elements right from the beginning—in charge of the production technology necessary and the logistical oversight of a multi-million-dollar project, very often as the only representative of the managerial corps present in Glashütte. His contributions were extremely important to the success of the A. Lange & Söhne brand.

Meanwhile, Knothe has been coordinating the ambitious expansion plans which will make this young company with its long tradition fit for the future. "A team of almost 400 employees is not the end of our development. We have a whole host of concepts in the top drawer that have literally accumulated over the years, which are continually modified by fine tuning from our team and feedback from the markets. Now we have entered a very exciting developmental phase, but we urgently need more production capacity, more qualified personnel, and fresh ideas. We are continually searching for engineers, technicians, and watchmakers. And," he adds, "we are maintaining the support and further development of our present team, practically as the foundation of our manufacture. Our experienced movement designers such as Helmut Geyer and Annegret Fleischer as well as our accomplished watchmakers like Udo Rudolf, Andreas Gelbrich, and Thomas Gemmel can't do it all. They can't develop new products and maintain old ones at the same time. Our collection continues to become more demanding, bringing more complexity with every new model."

Lange maintains its own school of watchmaking and additionally supports up-and-coming scientific talent at the Technical University as well as the College for Economics and Technology in Dresden. At the TU, students and doctoral candidates are supported who specially work in the pure research of watch movement technology. This research has already had an enormous influence on the repeated precision in the production and regulation of watch movements. "Bringing out new, interesting, and complicated watches and manufacturing them in-house with hundred-percent precision and perfect quality is, on the technical side, possibly one of the greatest challenges for A. Lange & Söhne," Knothe says. "For this is exactly what it is all about: not production numbers, but exclusivity, precision, and quality." ∎

A. Lange & Söhne's Success

Around the World

The success of the new A. Lange & Söhne collection didn't come as a surprise—the powers behind the project would hardly have embarked upon the venture if they weren't completely convinced of their concept. It was the waves that the once and now leading German brand's comeback caused that no one had foreseen.

The standard that had been set was very high when Lange began its frontal attack on the established luxury watch market. No less than the largest and most reputable of Switzerland's luxury brands and noble manufactures—including such illustrious names as Patek Philippe, Vacheron Constantin, Audemars Piguet, and Breguet—had been chosen as measuring sticks, and at an unusually unfavorable time at that.

At the beginning of the 1990s the market for high-quality, complicated mechanical wristwatches was experiencing a dampening effect after an unexpectedly high level of popularity no one had ever dreamed possible. Alongside established Genevan luxury brands, many other manufacturers, some with complicated movements purchased from the outside or extravagantly decorated jewelry watches, attempted to gain access to the peak of the price pyramid—some successfully, others less so. At any rate, this market segment was proportionally overcrowded in the mid-nineties, production capacity was overtaxed, and the myriad of products available could only be termed overwhelming. Retailers' inventory was overflowing, and under normal circumstances these watch and jewelry specialists, suffering from a sudden drop in business, would never have made the effort to bring in yet another watch collection. Under normal circumstances,

The first thirty-six sets of the Lange 1 Time Zone were simultaneously sent to thirty-six dealers around the world on July 6, 2005. Thirty-six watchmakers delivered the watches personally, observed by cameras transmitting the whole thing by satellite back to Glashütte and around the world live.

that is, but the launch of the first Lange watch collection in more than fifty years was anything but normal.

Excited by the effective and stylish presentation staged by the young, old brand in October 1994, and completely convinced of the watches' quality after affording them a first glance, numerous candidates stormed Lange Uhren GmbH before that year's Christmas season began, wanting to become official Lange dealers.

Günter Blümlein, who had experienced IWC and Jaeger-LeCoultre's "hard years" at the front line during the early 1980s, had to get used to this unusual situation: Lange did not need to search for dealers, instead the company had the privilege of choosing them.

Meteoric Ascent

It didn't take a fortune teller to predict that, at least in Germany, this luxury brand born again from literal ruins would experience a successful start. Enthusiasm for the newcomer from the East gripped not only passionate Lange pocket watch collectors, hyper-proud Germans, and homesick Saxons, but also ambitious watch collectors from every discipline and, above all, from every social stratum—including those who had never before wasted a thought on buying themselves a special (or expensive) watch.

Least expected in Glashütte was the positive resonance experienced in international circles, especially Switzerland, the generally accredited home of luxury wristwatches. Italians, who are rightly said to have a special feel for horological refinements and exclusive collectible objects, welcomed the descendants of Ferdinand Adolph Lange with open arms. And in the country possessing the largest number of watch collectors and enthusiasts worldwide, the USA, those in the know dashed to scoop up these mechanical gems made in Germany—even if they had a hard time getting their hands on them at first. Until the fall of 2002, A. Lange &

Lange at home in all time zones: At Beyer in Zürich and Simonetta Orsini in Buenos Aires (above), in Hong Kong at Carlson (with guest of honor Jackie Chan), at Sincere in Singapore, and at Wempe in Frankfurt (at left, from top to bottom), as well as live on a video screen in New York's Times Square (opposite).

Söhne's timepieces were only available at Cellini and Wempe in New York.

Lange's watches are available today in more than forty countries spanning the globe via 130 dealers with a good 190 points of sale. Not only is Europe well supplied with these watches, but North America and Asia also have a reasonable number of concessions. This is even more amazing when one stops to consider that the production capacity in Glashütte—despite a very committed increase in production room—is still kept exceptionally limited in terms of growth. A comparison with the industry giants of the luxury watch segment, for example, Rolex and Cartier, and even exclusive manufactures and noble brands such as Patek Philippe, Vacheron Constantin, and Audemars Piguet shows that they manufacture many more watches per year. Despite this, A. Lange & Söhne was already one of the "finest addresses" for important turnover among jewelers and watch specialty shops just a few years after its market introduction.

Although the company's list of authorized dealers reads like a *Who's Who* of international watch high society, the brand is under no circumstances over-distributed, either in its homeland or anywhere else. Lange is proud that, despite all temptation, it has been able to avoid this "most elementary of all mistakes in the watch industry," according to Günter Blümlein. Territory protection and contractually secured distribution relationships with obligations on both sides help avoid the danger of grey market activity in advance, and thus Lange's products enjoy excellent placement in the few windows in which they can be seen. Lange's business partners are understandably enthusiastic about showing off their best horse.

A Question of Communication

Without a doubt the A. Lange & Söhne brand profited above all in the initial phase from an advertising plan that was quite unusual for a niche product of this sort. Full-page or double-page spread advertising in widely circulated national magazines and newspapers caused quite a stir, and perfect PR supplied the media with solid information. Hardly a German-language publication could pass on editorial about the brand within the first twelve months of its watches' launch, setting a word-of-mouth propaganda machine in motion in "informed circles." Almost simultaneously the discussion surrounding the renaissance of the "emperor's

brand" spilled over into foreign countries, and in that electric field located between a not-quite digested fear of a reunited Teutonia and respect for the predicate "made in Germany," legends of precision, engineering art, and exclusivity flourished.

It was the products themselves that ensured that the actual topic "watches" wasn't forgotten in the midst of all of this attention. The timepieces' asking prices may have been high, but the perfectly wrapped package including credible "inner values" and an aesthetically pleasing appearance justified this positioning. Trade magazines couldn't avoid taking their hats off to the technical finesse of the large date, the tourbillon with its chain and fusee, and the automatically resetting second hand (zero reset).

A. Lange & Söhne's Success

Represented all over the world: at Wempe in Düsseldorf, Huber in Munich (above), King's Sign in Taipei, and Pendulum in Bangkok (below).

And the lovingly celebrated allusions to good old pocket watch tradition in the form of a three-quarter plate, gold chatons, and screw balance awakened emotions thought long gone. Those with a heart for watchmaking couldn't help loving these watches that subtly played with those emotions: They emanate a beautifully timeless charm, although they are progressive and innovative in every technological and aesthetic detail.

Sale of the Crown Jewels

Not quite ten years after being founded under the roof of Mannesmann/VDO, Lange Uhren GmbH came to new ownership: In the summer of 2000, the erstwhile united companies of LMH Holding (Les Manufactures Horlogères) Lange, Jaeger-LeCoultre, and IWC were sold to the Richemont Luxury Group for a record-breaking sum of 2.8 billion Swiss francs (approximately two billion American dollars).

With the integration of the LMH brands into the Richemont concern's portfolio, a number of wonderful opportunities for the future developed for A. Lange & Söhne along with its sibling brands IWC and the Jaeger-LeCoultre *manufacture* among the other watch brands already nestled there. These brands seem to be separated into two large groups: On one side there is Cartier, the largest brand with the most turnover within the mixed luxury concern, together with Baume & Mercier and the watch divisions of the brands Montblanc, Van

Cleef & Arpels, and Dunhill. On the other side, one previously found the "watchmaker" brands, Piaget, Vacheron Constantin, and Officine Panerai. These have now been complemented with the addition of A. Lange & Söhne, IWC, and Jaeger-LeCoultre.

Organizing the cooperation between the two different company cultures of Mannesmann and Richemont and joining both German and Swiss watchmaking elements should have been Günter Blümlein's largest and most honorable task. However, the charismatic manager and visionary marketing strategist passed away in October 2001, long before he could fulfill the task to his satisfaction—and, indeed, long before his time.

Showered with Awards

The overpoweringly clear choice of the Lange 1 model as the Watch of the Year 1995 by the readers of the German magazine *ArmbandUhren* was only the beginning of a long line of awards that the modern Lange watches, their creative fathers, and the entire A. Lange & Söhne project were to receive in the ensuing years. Popular and jury-decided competitions in Austria, Italy, and Japan paid tribute to the Lange 1 and other models. In order to avoid upsetting the local watch industry, a Swiss jury even created an extra category for the successful German watch—supposedly not in competition—in order to bestow honor upon it. The Lange advertising campaign was awarded a silver Effie in 1996, and then-German president Roman Herzog honored Lange for its courageous activity in the new German states in 1997.

In December 2000 the next German president, Johannes Rau, went to Glashütte to celebrate the tenth anniversary of the refounded watch manufacture with its employees and honored guests, finding appreciative words for the company's exemplary development. The emotional and very personal speech of the deceased statesman will always remain in the memories of those who attended.

Günter Blümlein found that the best award A. Lange & Söhne could receive was the increased general interest in watches originating in Glashütte: "Our entrepreneurial duty needs to be seen in light of the German reunification. Alongside economic calculations we always carried the emotional vision of contributing something to the reconstruction of former East Germany, turning ourselves into a locomotive of sorts. The region surrounding Glashütte was always at the forefront, and when I see the other Glashütte brands' success where A. Lange & Söhne has sown the seeds, then I am filled with pride and happiness—proud of our own success, and happy for the others' success."

Fabian Krone has retained this dedication to Glashütte and its surrounding area, continuing to write this success story in his own hand – a new state of tradition. ■

Live: Fabian Krone and Hartmut Knothe emceed the "Time Zone Event" with the charming Desirée Nosbusch in Glashütte, transmitted all over the world via satellite.

The Collection of A. Lange & Söhne

Lange 1
101.021

Movement: mechanical with manual winding, Lange Caliber L901.0, diameter 30.4 mm, height 5.9 mm, 53 jewels, nine of which are screw-mounted in gold chatons; 21,600 vph, glucydur screw balance, stop-seconds, Nivarox balance spring with special terminal curve, swan-neck fine adjustment and patented beat regulation; twin serially operating spring barrels; power reserve 72 hours; Glashütte three-quarter plate, bridges with Glashütte ribbing

Functions: hours, minutes, subsidiary seconds; outsize date (patented); power reserve display

Case: 18-karat yellow gold; diameter 38.5 mm, height 10 mm; sapphire crystal; case back secured with six screws and sapphire crystal exhibition window; corrector button for date display on edge of case

Dial: solid silver, champagne-colored; applied markers, Roman numerals and lance hands in yellow gold; displays: off-center hours and minutes at 9 o'clock, outsize date (double aperture) at 1 o'clock, subsidiary seconds at 5 o'clock, power reserve display at 3 o'clock

Band: crocodile skin; buckle

Lange 1 "Soirée"
110.030

The Soirée version of the Lange 1 model illustrates that this straightforward timepiece can be more than a technical and cultural showpiece: It can also play the sensitive role of an understated piece of jewelry. The mother-of-pearl dial made of select raw material carries a delicate guilloché pattern that, in conjunction with its rhodium-plated numerals, markers, and window frames, gives distinct structure to the various displays and scales located there.

Lange 1
101.033

Movement: mechanical with manual winding, Lange Caliber L901.0, diameter 30.4 mm, height 5.9 mm, 53 jewels, nine of which are screw-mounted in gold chatons; 21,600 vph, glucydur screw balance, stop-seconds, Nivarox balance spring with special terminal curve, swan-neck fine adjustment and patented beat regulation; twin serially operating spring barrels; power reserve 72 hours; Glashütte three-quarter plate, bridges with Glashütte ribbing

Functions: hours, minutes, subsidiary seconds; outsize date (patented); power reserve display

Case: 18-karat red gold; diameter 38.5 mm, height 10 mm; sapphire crystal; case back secured with six screws and sapphire crystal exhibition window; corrector button for date display on edge of case

Dial: solid silver, slate grey; applied markers, Roman numerals and lance hands in red gold; displays: off-center hours and minutes at 9 o'clock, outsize date (double aperture) at 1 o'clock, subsidiary seconds at 5 o'clock, power reserve display at 3 o'clock

Band: crocodile skin; buckle

Lange 1 Moon Phase
109.032

Movement: mechanical with manual winding, Lange Caliber L901.5, diameter 30.4 mm, height 5.9 mm, 54 jewels, nine of which are screw-mounted in gold chatons; 21,600 vph, glucydur screw balance, stop-seconds, Nivarox balance spring with special terminal curve, swan-neck fine adjustment and patented beat regulation; twin serially operating spring barrels; power reserve 72 hours; Glashütte three-quarter plate, bridges with Glashütte ribbing

Functions: hours, minutes, subsidiary seconds; outsize date (patented); power reserve display; continuously rotating moon phase display

Case: 18-karat red gold; diameter 38.5 mm, height 10.4 mm; sapphire crystal; case back secured with six screws and sapphire crystal exhibition window; corrector button for date display on edge of case

Dial: solid silver, argenté (silvered); applied markers, Roman numerals and lance hands in red gold, hands inlaid with SuperLumiNova; displays: off-center hours and minutes at 9 o'clock, outsize date (double aperture) at 1 o'clock, subsidiary seconds and moon phase at 5 o'clock, power reserve display at 3 o'clock

Band: crocodile skin; buckle

Lange 1 Time Zone
116.021

Movement: mechanical with manual winding, Lange Caliber L031.1, diameter 34.1 mm, height 6.65 mm, 54 jewels, four of which are screw-mounted in gold chatons; 21,600 vph, glucydur screw balance, stop-seconds, Nivarox balance spring with special terminal curve, swan-neck fine adjustment and patented beat regulation; twin serially operating spring barrels; power reserve 72 hours; Glashütte three-quarter plate, bridges with Glashütte ribbing

Functions: hours, minutes, subsidiary seconds; second time zone; outsize date; power reserve display; home time with day/night indication, local time (hours, minutes) with day/night indication and reference city ring settable via button

Case: 18-karat yellow gold; diameter 41.9 mm, height 11 mm; sapphire crystal; case back secured with six screws and sapphire crystal exhibition window; two corrector buttons for date display (at 10 o'clock) and time zone (at 8 o'clock) on edge of case

Dial: solid silver, champagne-colored; applied markers, Roman numerals and lance hands in yellow gold; displays: off-center hours and minutes at 9 o'clock, outsize date (double aperture) at 1 o'clock, second time zone at 5 o'clock, subsidiary seconds at 7 o'clock, power reserve display at 3 o'clock, two day/night indications, ring around edge of dial featuring 24 reference city names

Band: crocodile skin; buckle

Lange 1 Time Zone
116.032

The characteristic asymmetrical dial arrangement of the Lange 1 lends itself beautifully to the display of a second time zone. The world time watch by A. Lange & Söhne–shown here in a red gold case–disposes of two dials and two pairs of hands that can be set independently of each other as well as a rotating ring featuring city names, each of which represents one of the places located within each of the 24 time zones that our world is divided into.
Additionally, this time zone watch features two day/night indicators whose miniscule hands always run synchronized to their respective time zones. The wearer can choose which of the two time displays should show home time and which local time.

Grand Lange 1
115.025

Movement: mechanical with manual winding, Lange Caliber L901.2, diameter 30.4 mm, height 5.9 mm, 53 jewels, nine of which are screw-mounted in gold chatons; 21,600 vph, glucydur screw balance, stop-seconds, Nivarox balance spring with special terminal curve, swan-neck fine adjustment and patented beat regulation; twin serially operating spring barrels; power reserve 72 hours; Glashütte three-quarter plate, bridges with Glashütte ribbing

Functions: hours, minutes, subsidiary seconds; outsize date (patented); power reserve display

Case: platinum; diameter 41.9 mm, height 11 mm; sapphire crystal; case back secured with six screws and sapphire crystal exhibition window; corrector button for date display on edge of case

Dial: solid silver, two-tone rhodium-plated/silver-plated; applied markers, Roman numerals and lance hands in white gold; displays: off-center hours and minutes at 9 o'clock, outsize date (double aperture) at 1 o'clock, subsidiary seconds and moon phase at 5 o'clock, power reserve display at 3 o'clock

Band: crocodile skin; buckle

Lange 1 Luminous
101.329

Movement: mechanical with manual winding, Lange Caliber L901.2, diameter 30.4 mm, height 5.9 mm, 53 jewels, nine of which are screw-mounted in gold chatons; 21,600 vph, glucydur screw balance, stop-seconds, Nivarox balance spring with special terminal curve, swan-neck fine adjustment and patented beat regulation; twin serially operating spring barrels; power reserve 72 hours; Glashütte three-quarter plate, bridges with Glashütte ribbing

Functions: hours, minutes, subsidiary seconds; outsize date (patented); power reserve display

Case: 18-karat white gold; diameter 38.5 mm, height 10 mm; sapphire crystal; case back secured with six screws and sapphire crystal exhibition window; corrector button for date display on edge of case

Dial: solid silver, black; Roman numerals and lance hands all inlaid with SuperLumiNova; displays: off-center hours and minutes at 9 o'clock, outsize date (double aperture) at 1 o'clock, subsidiary seconds at 5 o'clock, power reserve display at 3 o'clock

Band: white gold; double-folding clasp

The Collection

Saxonia
105.021

Movement: mechanical with manual winding, Lange Caliber L941.3, diameter 25.6 mm, height 4.95 mm, 30 jewels, four of which are screw-mounted in gold chatons; 21,600 vph, glucydur screw balance, stop-seconds, Nivarox balance spring with swan-neck fine adjustment; patented beat regulation via micrometer screw adjustment; power reserve 42 hours; Glashütte three-quarter plate, bridges with Glashütte ribbing

Functions: hours, minutes, subsidiary seconds; outsize date (patented)

Case: 18-karat yellow gold; diameter 33.9 mm, height 9.1 mm; sapphire crystal; case back secured with six screws and sapphire crystal exhibition window; corrector button for date display on edge of case

Dial: solid silver, champagne-colored; applied markers, Roman numerals and lance hands in yellow gold; displays: outsize date (double aperture) at 12 o'clock, subsidiary seconds at 6 o'clock

Band: crocodile skin; buckle

Saxonia
803.031

The Saxonia model was replaced by the Langematik as Lange's "simply elegant men's watch," answering a growing demand for timepieces somewhat larger in dimension. This has allowed the graceful Saxon watch some freedom to be outfitted a little more femininely, also functioning as a timekeeper for quality- and status-conscious women with its 34-mm diameter. The pictured white gold version with black dial additionally sports a seductive wreath of 48 baguette-cut diamonds of highest (top Wesselton) quality on the bezel for a total of 1.9 carats.

Cabaret
107.031

Movement: mechanical with manual winding, shaped movement Lange Caliber L931.3, dimensions 25.6 x 17.6 mm, height 4.95 mm, 30 jewels, three of which are screw-mounted in gold chatons; 21,600 vph, glucydur screw balance, stop-seconds, Nivarox balance spring with swan-neck fine adjustment; patented beat regulation with micrometer screw adjustment; power reserve 42 hours; Glashütte three-quarter plate, bridges with Glashütte ribbing

Functions: hours, minutes, subsidiary seconds; outsize date (patented)

Case: 18-karat red gold; dimensions 36.3 x 25.5 mm, height 9.1 mm; sapphire crystal; case back secured with six screws and sapphire crystal exhibition window; corrector button for date display on edge of case

Dial: solid silver, black; applied markers and Roman numerals in red gold, red gold lance hands; displays: outsize date (double aperture) at 12 o'clock, subsidiary seconds at 6 o'clock

Band: crocodile skin; buckle

Cabaret Moon Phase
118.021

Movement: mechanical with manual winding, shaped movement Lange Caliber L931.5, dimensions 25.6 x 17.6 mm, height 5.05 mm, 31 jewels, three of which are screw-mounted in gold chatons; 21,600 vph, glucydur screw balance, stop-seconds, Nivarox balance spring with swan-neck fine adjustment; patented beat regulation with micrometer screw adjustment; power reserve 42 hours; Glashütte three-quarter plate, bridges with Glashütte ribbing

Functions: hours, minutes, subsidiary seconds; outsize date (patented); moon phase display

Case: 18-karat yellow gold; dimensions 36.3 x 25.5 mm, height 9.1 mm; sapphire crystal; case back secured with six screws and sapphire crystal exhibition window; corrector button for date display on edge of case

Dial: solid silver, champagne-colored; applied markers and Roman numerals in yellow gold, golden lance hands; displays: outsize date (double aperture) at 12 o'clock, subsidiary seconds at 6 o'clock

Band: crocodile skin; buckle

Arkade
103.021

Movement: mechanical with manual winding, shaped movement Lange Caliber L911.4, dimensions 25.6 x 17.6 mm, height 4.95 mm, 30 jewels, three of which are in screw-mounted gold chatons; 21,600 vph, glucydur screw balance, stop-seconds, Nivarox balance spring with swan-neck fine adjustment; patented beat regulation via micrometer screw adjustment; power reserve 42 hours; Glashütte three-quarter plate, bridges with Glashütte ribbing

Functions: hours, minutes, subsidiary seconds; outsize date (patented)

Case: 18-karat yellow gold; dimensions 29 x 22.2 mm, height 8.4 mm; sapphire crystal; case back secured with four screws and sapphire crystal exhibition window; corrector button for date display on edge of case

Dial: solid silver, champagne-colored; applied markers and Roman numerals in yellow gold, golden lance hands; displays: outsize date (double aperture) at 12 o'clock, subsidiary seconds at 6 o'clock

Band: crocodile skin; buckle

Arkade
103.035

The platinum Arkade with black dial represents an unusual and subtly charming variation of this model. It sports a fairly unfeminine-sounding color and material combination on paper. In reality, however, this union turns out to be very attractive, the dial making a "sporty" contribution to the elegantly designed case, the shape of which is reminiscent of the Dresden castle's arcades.

Richard Lange
232.032

Movement: mechanical with manual winding, Lange Caliber L041.2, diameter 30.4 mm, height 6 mm, 27 jewels, two of which are in screw-mounted gold chatons; 21,600 vph, shockproofed glucydur screw balance with eccentric regulating screws, stop-seconds, in-house balance spring with spring clamp patent pending, patented beat regulation with micrometer screw adjustment and swan-neck spring; power reserve 38 hours; Glashütte three-quarter plate, bridges with Glashütte ribbing

Functions: hours, minutes, sweep seconds

Case: 18-karat red gold; diameter 40.5 mm, height 10.6 mm; sapphire crystal; case back secured with six screws and sapphire crystal exhibition window

Dial: solid silver, argenté (silvered); Roman numerals and minute scale, gold lance hands; blued second hand

Band: crocodile skin; buckle

Richard Lange
232.025

Movement: mechanical with manual winding, Lange Caliber L041.2, diameter 30.4 mm, height 6 mm, 27 jewels, two of which are in screw-mounted gold chatons; 21,600 vph, shockproofed glucydur screw balance with eccentric regulating screws, stop-seconds, in-house balance spring with spring clamp patent pending, patented beat regulation with micrometer screw adjustment and swan-neck spring; power reserve 38 hours; Glashütte three-quarter plate, bridges with Glashütte ribbing

Functions: hours, minutes, sweep seconds

Case: platinum; diameter 40.5 mm, height 10.6 mm; sapphire crystal; case back secured with six screws and sapphire crystal exhibition window

Dial: solid silver, rhodium-plated; Roman numerals and minute scale, gold lance hands; blued second hand

Band: crocodile skin; buckle

The Collection

1815
206.021

Movement: mechanical with manual winding, Lange Caliber L941.1, diameter 25.6 mm, height 3.2 mm, 21 jewels, four of which are in screw-mounted gold chatons; 21,600 vph, glucydur screw balance, stop-seconds, Nivarox balance spring with swan-neck fine adjustment, patented beat regulation with micrometer screw adjustment; power reserve 45 hours; Glashütte three-quarter plate, bridges with Glashütte ribbing

Functions: hours, minutes, subsidiary seconds

Case: 18-karat yellow gold; diameter 35.9 mm, height 7.5 mm; sapphire crystal; case back secured with six screws and sapphire crystal exhibition window

Dial: solid silver, argenté (silvered); Arabic numerals and minute scale, blued lance hands; displays: subsidiary seconds at 6 o'clock

Band: crocodile skin; buckle

1815 UP and DOWN
221.025

Movement: mechanical with manual winding, Lange Caliber L942.1, diameter 25.6 mm, height 3.7 mm, 27 jewels, six of which are in screw-mounted gold chatons; 21,600 vph, glucydur screw balance, stop-seconds, Nivarox balance spring with swan-neck fine adjustment, patented beat regulation with micrometer screw adjustment; power reserve 45 hours; Glashütte three-quarter plate, bridges with Glashütte ribbing

Functions: hours, minutes, subsidiary seconds; power reserve display

Case: platinum; diameter 35.9 mm, height 7.9 mm; sapphire crystal; case back secured with six screws and sapphire crystal exhibition window

Dial: solid silver, argenté (silvered); Arabic numerals and minute scale, blued lance hands; displays: subsidiary seconds at 4 o'clock, power reserve display at 8 o'clock

Band: crocodile skin; buckle

1815 Automatic
303.332

Movement: mechanical with automatic winding, Lange Caliber L921.2 SAX-O-MAT, diameter 30.4 mm, height 3.8 mm, 36 jewels; 21,600 vph, glucydur screw balance, Nivarox balance spring with swan-neck fine adjustment, patented beat regulation with micrometer screw adjustment; bilaterally winding embossed three-quarter rotor in 21-karat gold and platinum with four ball bearings with an alternating gear; power reserve 46 hours; bridges with Glashütte ribbing

Functions: hours, minutes, subsidiary seconds; stop-seconds with automatic zero-reset function of the second hand when the crown is pulled

Case: 18-karat red gold; diameter 37 mm, height 8.2 mm; sapphire crystal; case back secured with six screws and sapphire crystal exhibition window

Dial: solid silver, argenté (silvered); Arabic numerals and minute scale, blued lance hands; displays: subsidiary seconds at 6 o'clock

Band: red gold; double-folding clasp

1815 Automatic
303.025

Movement: mechanical with automatic winding, Lange Caliber L921.2 SAX-O-MAT, diameter 30.4 mm, height 3.8 mm, 36 jewels; 21,600 vph, glucydur screw balance, Nivarox balance spring with swan-neck fine adjustment, patented beat regulation with micrometer screw adjustment; bidirectionally winding embossed three-quarter rotor in 21-karat gold and platinum with four ball bearings on the reverser; power reserve 46 hours; bridges with Glashütte ribbing

Functions: hours, minutes, subsidiary seconds; stop-seconds with automatic zero-reset function of the second hand when the crown is pulled

Case: platinum; diameter 37 mm, height 8.2 mm; sapphire crystal; case back secured with six screws and sapphire crystal exhibition window

Dial: solid silver, argenté; Arabic numerals and minute scale, blued lance hands; displays: subsidiary seconds at 6 o'clock

Band: crocodile skin; buckle

1815 Chronograph
401.031

Movement: mechanical with manual winding, Lange Caliber L951.0, diameter 30.6 mm, height 6.1 mm, 34 jewels, four of which are screw-mounted in gold chatons; 18,000 vph, glucydur screw balance, stop-seconds, patented beat regulation, Nivarox balance spring, control of chronograph functions with column wheel; Glashütte three-quarter plate, bridges with Glashütte ribbing; movement decorated and engraved by hand

Functions: hours, minutes, subsidiary seconds; chronograph with flyback function and precisely jumping minute counter

Case: 18-karat red gold; diameter 39 mm, height 10.8 mm; sapphire crystal; case back secured with six screws and sapphire crystal exhibition window

Dial: solid silver, black; Arabic numerals, pulsometer and minute scales, red gold lance hands; displays: subsidiary seconds at 8 o'clock, 30-minute counter at 4 o'clock, chronograph sweep second hand

Band: crocodile skin; buckle

1815 Chronograph
401.326

Movement: mechanical with manual winding, Lange Caliber L951.0, diameter 30.6 mm, height 6.1 mm, 34 jewels, four of which are screw-mounted in gold chatons; 18,000 vph, glucydur screw balance, stop-seconds, patented beat regulation, Nivarox balance spring, control of chronograph functions with column wheel; Glashütte three-quarter plate, bridges with Glashütte ribbing; movement decorated and engraved by hand

Functions: hours, minutes, subsidiary seconds; chronograph with flyback function and precisely jumping minute counter

Case: 18-karat white gold; diameter 39 mm, height 10.8 mm; sapphire crystal; case back secured with six screws and sapphire crystal exhibition window

Dial: solid silver, argenté (silvered); Arabic numerals, pulsometer and minute scales, blued lance hands; displays: subsidiary seconds at 8 o'clock, 30-minute counter at 4 o'clock, chronograph sweep second hand

Band: white gold; double-folding clasp

Tourbograph Pour le Mérite
702.025

Split-seconds chronograph Caliber L903.0, comprising 465 individual components, is driven by a chain-and-fusee transmission and also features a power reserve display and a tourbillon. Its plates and bridges are crafted in German silver; the chronograph bridge is engraved by hand. The 8.9 mm high movement has a diameter of 30 mm and is outfitted with 43 jewels, two of which are diamond endstones for the tourbillon. A diamond is set into the polished setting lever for the chronograph clutch. It serves as an endstone to limit the tourbillon's endshake.

Tourbograph Pour le Mérite
702.025

Movement: mechanical with manual winding, Lange Caliber L903.0, diameter 30 mm, height 8.9 mm, 43 jewels, six of which are screw-mounted in gold chatons, tourbillon cage bearing with two diamond endstones; 21,600 vph, glucydur screw balance (with shock protection), one-minute tourbillon, Nivarox balance spring, constant force regulation with chain and fusee transmission as well as stepped planetary gear; power reserve 36 hours; Glashütte three-quarter plate, bridges with Glashütte ribbing

Functions: hours, minutes; split-seconds chronograph; power reserve display

Case: platinum; diameter 41.2 mm, height 14.3 mm; sapphire crystal; case back secured with six screws and sapphire crystal exhibition window; buttons for sweep chronograph, zero reset and rattrapante

Dial: solid silver, argenté (silvered); Arabic numerals and minute scale, blued-steel lance hands; cutaway at 6 o'clock for baton-shaped bridge above the tourbillon cage; displays: power reserve at 3 o'clock; sweep chronograph and rattrapante hands

Band: crocodile skin; buckle

Limited edition of 51 in platinum and 50 in other metals.

The Collection

Langematik
308.021

Movement: mechanical with automatic winding, Lange Caliber L921.4 SAX-O-MAT, diameter 30.4 mm, height 5.55 mm, 45 jewels; 21,600 vph, glucydur screw balance, Nivarox balance spring with swan-neck fine adjustment; patented beat regulation with micrometer screw adjustment; bidirectionally rotating embossed three-quarter rotor in 21-karat gold and platinum with four ball bearings on the reverser; power reserve 46 hours; bridges with Glashütte ribbing

Functions: hours, minutes, subsidiary seconds; outsize date (patented); stop-seconds with automatic resetting device (zero reset) when the crown is pulled out

Case: 18-karat yellow gold; diameter 37 mm, height 9.7 mm; sapphire crystal; case back secured with six screws and sapphire crystal exhibition window; corrector button for date display on edge of case

Dial: solid silver, champagne-colored; applied markers and lance hands in yellow gold with SuperLumiNova inlay; displays: outsize date (double aperture) at 12 o'clock, subsidiary seconds at 6 o'clock

Band: crocodile skin; buckle

Langematik
308.027

Movement: mechanical with automatic winding, Lange Caliber L921.4 SAX-O-MAT, diameter 30.4 mm, height 5.55 mm, 45 jewels; 21,600 vph, glucydur screw balance, Nivarox balance spring with swan-neck fine adjustment; patented beat regulation with micrometer screw adjustment; bidirectionally rotating embossed three-quarter rotor in 21-karat gold and platinum with four ball bearings on the reverser; power reserve 46 hours; bridges with Glashütte ribbing

Functions: hours, minutes, subsidiary seconds; outsize date (patented); stop-seconds with automatic resetting device (zero reset) when the crown is pulled out

Case: 18-karat white gold; diameter 37 mm, height 9.7 mm; sapphire crystal; case back secured with six screws and sapphire crystal exhibition window; corrector button for date display on edge of case

Dial: solid silver, black; applied markers and lance hands in rhodium-plated gold with SuperLumiNova inlay; displays: outsize date (double aperture) at 12 o'clock, subsidiary seconds at 6 o'clock

Band: crocodile skin; buckle

Langematik Perpetual
310.021

Movement: mechanical with automatic winding, Lange Caliber L922.1 SAX-O-MAT, diameter 30.4 mm, height 5.7 mm, 43 jewels; 21,600 vph, glucydur screw balance, Nivarox balance spring with swan-neck fine adjustment; patented beat regulation with micrometer screw adjustment; bidirectionally rotating embossed three-quarter rotor in 21-karat gold and platinum with four ball bearings on the reverser; power reserve 46 hours; bridges with Glashütte ribbing

Functions: hours, minutes, subsidiary seconds; perpetual calendar (outsize date, day of the week, month, moon phase, leap year); 24-hour display with day/night indication; stop-seconds with automatic resetting device (zero reset) when the crown is pulled out

Case: 18-karat yellow gold; diameter 38.5 mm, height 10.2 mm; sapphire crystal; case back secured with six screws and sapphire crystal exhibition window; synchronized correction buttons on edge of case

Dial: solid silver, champagne-colored; applied markers and lance hands in yellow gold with SuperLumiNova inlay; displays: outsize date (double aperture) at 12 o'clock, day of the week and 24-hour display at 9 o'clock, subsidiary seconds and moon phase at 6 o'clock, leap year indication at 4 o'clock, month at 3 o'clock

Band: crocodile skin; buckle

Langematik Perpetual
310.225

In the next 94 years, the first Lange wristwatch featuring a perpetual calendar will automatically take the varying lengths of the months into consideration. For this reason the wearer only needs to correct the display by one day on March 1, 2100 as February 29 will not exist in that year. Since the current owner of a Langematik Perpetual most likely will not be experiencing this day, the next generation will be very pleased about its uncomplicated correction process: Next to the individual correctors for various displays recessed into the case, this timepiece disposes of another practical recessed button with which all five calendar displays – day, month, year, leap year, and moon phase – can be adjusted in a synchronized manner. The platinum version pictured here features an integrated platinum bracelet with shaped lugs and a double folding clasp.

Datograph
403.032

Movement: mechanical with manual winding, Lange Caliber L951.1, diameter 30.6 mm, height 7.5 mm, 40 jewels, four of which are screw-mounted in gold chatons; 18,000 vph, glucydur screw balance, stop-seconds, patented beat regulation; Nivarox balance spring, column wheel control of chronograph functions; Glashütte three-quarter plate, bridges with Glashütte ribbing; movement decorated and engraved by hand

Functions: hours, minutes, subsidiary seconds; outsize date (patented); chronograph with flyback function and exactly jumping minute counter

Case: 18-karat red gold; diameter 39 mm, height 12.8 mm; sapphire crystal; case back secured with six screws and sapphire crystal exhibition window; correction button for date display on edge of case

Dial: solid silver, black; applied Roman numerals and lance hands in red gold with SuperLumiNova inlay; displays: outsize date (double aperture) at 12 o'clock, subsidiary seconds at 8 o'clock, 30-minute counter at 4 o'clock, sweep chronograph second hand

Band: crocodile skin; buckle

Datograph
403.035

A. Lange & Söhne presented the first chronograph in the company's history of wristwatches at the brand at the Basel Fair in 1999. The brand would never have been able to do justice to its own high demands if the movement designers had not thought up a few new and unusual solutions for this model, available in limited numbers. Thus, the Datograph, along with a column-wheel controlled chronograph with flyback function for stopping times in succession without having to first reset, also features the patented Lange outsize date. Explaining where the clever name of the timepiece comes from—"date" and "chronograph."

Lange Double Split
404.035

Movement: mechanical with manual winding, Lange Caliber L001.1, diameter 30.6 mm, height 9.45 mm, 40 jewels, four of which are screw-mounted in gold chatons; 21,600 vph, glucydur screw balance with shock protection and an in-house balance spring with balance spring clamp registered for a patent, stop-seconds, chronograph functions controlled by two column wheels, isolator mechanism; Glashütte three-quarter plate, bridges with Glashütte ribbing; movement decorated and engraved by hand

Functions: hours, minutes, subsidiary seconds; split-seconds chronograph for second and minute counters (precisely jumping); power reserve display

Case: platinum; diameter 43,2 mm, height 15.3 mm; sapphire crystal; case back secured with six screws and sapphire crystal exhibition window

Dial: solid silver, black; rhodium-plated, applied Roman numerals and lance hands in gold with SuperLumiNova inlay; displays: power reserve display at 12 o'clock, subsidiary seconds at 8 o'clock, 30-minute counter at 4 o'clock, sweep chronograph and rattrapante second hands

Band: crocodile skin; buckle

Datograph Perpetual
410.025

Movement: mechanical with manual winding, Lange Caliber L952.1, diameter 32 mm, height 8 mm, 45 jewels, four of which are screw-mounted in gold chatons; 18,000 vph, glucydur screw balance with shock protection and in-house balance spring with balance spring clamp registered for a patent, stop-seconds; column wheel control of chronograph functions; Glashütte three-quarter plate; bridges with Glashütte ribbing; movement decorated and engraved by hand

Functions: hours, minutes, subsidiary seconds; perpetual calendar (outsize date, day of the week, month, moon phase, leap year); day/night indication; chronograph with flyback function and precisely jumping minute counter

Case: platinum; diameter 41 mm, height 13.5 mm; sapphire crystal; case back secured with six screws and sapphire crystal exhibition window; correction button for date display on edge of case

Dial: solid silver, rhodium-plated; minute and pulsometer scales, applied Roman numerals and lance hands in gold with SuperLumiNova inlay and rhodium-plated; displays: outsize date (double aperture) at 12 o'clock, subsidiary seconds and day of the week as well as day/night indication at 8 o'clock, 30-minute counter, month and leap year indication at 4 o'clock, moon phase at 6 o'clock, sweep chronograph second hand

Band: crocodile skin; buckle

A Word about the Third Edition

The first five years after the acclaimed unveiling of A. Lange & Söhne's debut collection on October 24, 1994 were already characterized by an incredibly rapid development of the brand and its collection—reason enough for us to dedicate our first *ArmbandUhren* special to this new star in the horological sky, which was brought to market just in time for the Swiss spring fairs in 1999. Due to the diligent creativity of these Saxon watchmakers, a first update was already necessary three short years later, and now, in 2006, we have put together a completely reworked and substantially extended publication depicting the newest developments in personnel, structure, and collection.

More than fifteen years have passed since the refounding of A. Lange & Söhne, and to simply state that a lot has happened in Glashütte would definitely be to greatly oversimplify the matter.

The individual model families have been arranged here in a new manner and illustrated with up-to-date photography. A number of models have already sold out since the first edition of this book, now making them among the most sought-after collector's pieces on the market. Since they are, however, still part of this young legend, we haven't cut them out completely but left them in their respective collections, marked accordingly.

A great deal of effort flowed into the completely new design of the chapter regarding the manufacture and its people, and we would like to especially thank Kirsten Hultzsch for her committed organizational aid in the tightrope act between photo shoots, interviews, and fact-checking. Photographer Jürgen Jeibmann took advantage of his location and nipped over to the manufacture several times to get just the right shot—before breakfast, on his lunch break, and even in the evenings. Getting all the players together in one spot on one day was simply an impossibility.

Thanks are also due to all of the Lange employees, who were committed in aiding us during the entire project all the way into the "hot" phase of end production, so that we can truly say that everyone really added his or her two cents to the success of this *ArmbandUhren* special.

Peter Braun

The *ArmbandUhren* publication A. Lange & Söhne is a special issue of the magazine *ArmbandUhren* and was published by

HEEL Verlag GmbH
Gut Pottscheidt
53639 Königswinter
Germany
Tel.: +49 2223 9230-0
Fax: +49 2223 9230-26
www.heel-verlag.de
info@heel-verlag.de

© 2006 HEEL Verlag GmbH, Königswinter

Editor-in-Chief:
Peter Braun

Authors:
Gerhard Claußen, Manfred Fritz, Peter Braun

English Text:
Elizabeth Doerr

Copyediting:
Ashley Benning

Photographs:
Lange Uhren GmbH, Jürgen Jeibmann, Holger Krull, Dr. Crott Auctions, Peter Braun, *ArmbandUhren* archives, Bertram Kober/PUNCTUM, oil paintings by Steffen Imhof, Isolde Fischer, Saxon State Library – State and University Library Dresden, German Photographic Collection Department

Cover and Layout:
Grafikbüro Schumacher, Königswinter

Color Separation:
Collibri Prepress, Königswinter

Printing:
Koelblin-Fortuna-Druck, Baden-Baden

Printed in Germany

– All Rights Reserved –

ISBN-10: 3-89880-624-3